Macromolecular Symposia 221

Fillers for Formulations

Alicante, Spain
September 8–11, 2003

Symposium Editor:
J. M. Martín-Martínez, Alicante, Spain

pp. 1–280 · January 2005
ISBN 3-527-31324-9

WILEY-VCH

Macromolecular Symposia publishes lectures given at international symposia and is issued irregularly, with normally 14 volumes published per year. For each symposium volume, an Editor is appointed. The articles are peer-reviewed. The journal is produced by photo-offset lithography directly from the authors' typescripts.
Further information for authors can be found at http://www.ms-journal.de
Suggestions or proposals for conferences or symposia to be covered in this series should also be sent to the Editorial office (E-mail: macro-symp@wiley-vch.de).

Macromolecular Symposia:
Annual subscription rates 2005
Macromolecular Full Package: including Macromolecular Chemistry & Physics (24 issues), Macromolecular Rapid Communications (24), Macromolecular Bioscience (12), Macromolecular Theory & Simulations (9), Macromolecular Materials and Engineering (12), Macromolecular Symposia (14):

Europe	Euro	7.088 / 7.797
Switzerland	Sfr	12.448 / 13.693
All other areas	US$	8.898 / 9.788

print only **or** electronic only / print **and** electronic

Postage and handling charges included. All Wiley-VCH prices are exclusive of VAT. Prices are subject to change.

Single issues and back copies are available. Please ask for details at: service@wiley-vch.de

Orders may be placed through your bookseller or directly at the publishers:
WILEY-VCH Verlag GmbH & Co. KGaA, P. O. Box 10 11 61, 69451 Weinheim, Germany, Tel. +49 (0) 62 01/6 06-400, Fax +49 (0) 62 01/60 61 84, E-mail: service@wiley-vch.de

For **copyright permission** please contact Claudia Jerke at:
Fax: +49 (0) 62 01/6 06-332, E-mail: cjerke@wiley-vch.de

For USA and Canada: Macromolecular Symposia (ISSN 1022-1360) is published with 14 volumes per year by WILEY-VCH Verlag GmbH & Co. KGaA, Boschstr. 12, 69451 Weinheim, Germany. Air freight and mailing in the USA by Publications Expediting Inc., 200 Meacham Ave., Elmont, NY 11003, USA. Application to mail at Periodicals Postage rate is pending at Jamaica, NY 11431, USA. POSTMASTER please send address changes to: Macromolecular Symposia, c/o Wiley-VCH, III River Street, Hoboken, NJ 07030, USA.

Printing: Strauss Offsetdruck, Mörlenbach. Binding: J. Schäffer, Grünstadt

Macromolecular Symposia

Articles published on the web will appear several weeks before the print edition. They are available through:

WILEY InterScience®

www.ms-journal.de

www.interscience.wiley.com

Eurofillers 2003 Conference
Alicante (Spain), 2003

Fillers for Polymeric Nanocomposites

Silica Fillers

Fillers as Flame Retardants

Modelling in Filled Polymers

Recycling of Fillers

Miscellaneous Aspects of Fillers

Author Index

Preface

The state of the art and advances in fillers' manufacturing, characterization and applications are specifically considered every two years at Eurofillers Conferences. Eurofillers 2003 was the first conference devoted to fillers organized in Spain, and it was mainly focussed on the analysis and understanding of filler–polymer interactions and their implications in several different industrial applications.

Several people and organizations actively participated in the preparation of Eurofillers 2003. Furthermore, the members of the Scientific and Organizing Committees took care of the scientific content of the conference, and the members of the Adhesion and Adhesives Laboratory of the University of Alicante organized the local arrangements of the conference.

About 170 delegates from 33 different countries from Europe, Africa, Asia, America and Australia participated in the Eurofillers 2003 Conference. 51 oral and 58 poster presentations were given.

There is no doubt of the importance of the understanding of the physics and chemistry of fillers to optimise the performance of filled plastics, polymers, rubbers and adhesives, mainly in the field of the nanotechnology. This book provides an interesting contribution to the existing literature by adding to the current less explored knowledge of fillers and their implications in formulations. Most of the contributions compiled in this book concern to the filler–polymer interactions mainly for nanocomposites, and some contributions are devoted to the incidence of fillers in adhesion. The papers in this book were organized into seven subheadings to facilitate searching and compile papers with similar information: adhesion, nanocomposites, silica fillers, flame retardants, recycling, modelling and miscellaneous. I believe that the papers have a high standard and I hope that they will remain as reference publications for all those interested in filler–polymer interactions and fillers performance in polymeric systems.

Obviously, more aspects need to be understood about the design and application of fillers, mainly as additives for polymers. There will be a lot of fascinating problems to discuss during the next Eurofillers 2005 Conference to be held in Bruges (Belgium) on 9–12 May, 2005 (Chairmen: P. Dubois, G. Groeninckx, R. Jérôme, and R. Legras).

J. M. Martín-Martínez

Interactions in Nanosilica-Polyurethane Composites Evidenced by Plate-Plate Rheology and DMTA

Virtudes Navarro-Bañón,[1] *José Vega-Baudrit,**[2] *Patricia Vázquez,*[3] *José Miguel Martín-Martínez*[1]

[1] Adhesion and Adhesives Laboratory, Department of Inorganic Chemistry, University of Alicante, 03080 Alicante, Spain
[2] Laboratorio de Polímeros – POLIUNA, Universidad Nacional, Apartado 86-3000, Heredia, Costa Rica
E-mail: jvegab@una.ac.cr
[3] Centro de Investigación y Desarrollo en Ciencias Aplicadas -CINDECA, CONICET-UNLP, 47 n 257 (B1900AJK) La Plata, Buenos Aires, Argentina

Summary: Fumed nanosilicas with different degree of silanization were used to obtain nanosilica-polyurethane (TPU) composites to modify their rheological and mechanical properties. The higher the silanol content of the nanosilica, the higher the storage modulus and the thermal stability of the silica-TPU composite. DMTA experiments confirmed that the TPU-silica interactions increased by increasing the silanol content of the nanosilica.

Keywords: adhesion; adhesive; nanosilica; polyurethane; rheology

Introduction

Thermoplastic polyurethanes (TPU) are fully reacted segmented block polymers having good mechanical and elastic properties. TPUs are produced by reaction of an isocyanate, a polyol and a chain extender, and they contain urethane moieties. The two-phase microstructure of the TPU is due to incompatibility between the soft and the hard segments. [1] The hard rigid segments (produced by reaction of the isocyanate and the chain extender) segregate into a glassy or semicrystalline domain, and the polyol soft segments form amorphous rubbery matrices in which the hard segments are dispersed.

Fumed silica nanoparticles (nanosilicas) are commonly added to improve the rheological and mechanical properties of TPU adhesives. [2-5] This improvement has been ascribed to the creation of hydrogen bonds between the silanol groups on the nanosilica surface and the urethane groups in the soft segments. To confirm this hypothesis, in this study a full silanated nanosilica was treated with a silane by Wacker-Chemie (Burghausen, Germany) to produce different nanosilicas with decreased silanol content. Then, different TPU composites containing 10 wt% nanosilicas with different degree of silanization were prepared and characterized using plate-plate rheometry and DMTA.

 DOI: 10.1002/masy.200550301

Experimental

Fumed silicas (nanosilicas) were manufactured by Wacker-Chemie (Burghausen, Germany) for this study. The degree of silanization was varied between 15.5 and 100 % by treatment with a silane (Di-Me-Di). Primary particle size was 7 nm in all nanosilicas. Table 1 shows some properties of the nanosilicas provided by Wacker-Chemie.

Table 1. Some properties of the nanosilicas.[1]

Nanosilica	SiOH (%)	Carbon content (%)	mmol SiOH/g silica
N20	100	0	0.60
H20	57.5	1.2	0.34
H20RD	15.5	4.7	0.09

[1] Data provided by Wacker-Chemie.

The specific surface area of the nanosilicas was obtained from N_2/77K adsorption isotherms measured in a Quantachrome adsorption system. The nanosilicas were outgassed at 100°C for 8 h at 10^{-6} torr before adsorption isotherms were obtained. On the other hand, the wettability of the nanosilicas was obtained as a measure of their hydrophilicity by placing them in deionized bidistilled water, ethanol, and ethanol/deionized bidistilled water mixture. The IR spectra of the nanosilicas were obtained in a FTIR Bruker IFS 66 system equipped with a MTEC 300 photoacoustic cell (Institut Fertgungstechnik Materialforschung – IFAM, Germany) and transmission electron microscopy – TEM - was used to analyze the degree of aggregation of the nanosilicas in a JEOL JEM-2010 instrument by using an acceleration voltage of 100kV.
TPU was synthesized by reacting 4,4-diphenyl methane diisocyanate –MDI-, polyadipate of 1,4-butanediol (Hoopol F-530, Hoocker) and 1,4-butanediol. A hard/soft segment (NCO/OH) ratio of 1.05 was used. 10 wt% (based on the polymer) nanosilica was added to 18 wt% TPU in MEK solution and stirred at 2500 rpm for 3 hours to produce a good dispersion of the nanosilica in the polymer. About 100 ml adhesive solution was placed in a teflon mould which was placed into a desiccator fully saturated of MEK vapour, to allows a slow evaporation of MEK and avoid bubbles formation in the TPU-nanosilicas composites. One week after, the mould was placed under open air for 3 days to allow complete evaporation of MEK, and thus the TPU-nanosilica composites were obtained. The nomenclature of the composites used in this study was PU0 (TPU without silica), and PU15.5, PU57.5 and PU100 (TPUs containing 15.5, 57.5 and 100 % silanol groups, respectively).

The rheological properties of the TPUs were characterized in a Bohlin CS50 stress controlled rheometer using a plate-plate (20 mm diameter) geometry. The experiments were carried out in oscillatory mode at a frequency of 1 Hz to determine the storage and loss modulus as a function of the temperature. The experiments were carried out from 200°C to 30°C, using a cooling rate of 5 °C/min. Furthermore, dynamic mechanical thermal analysis (DMTA) was performed in a Rheometric Scientific DMTA Mk III instrument using the two-point bending mode. The experiments were carried out from –80°C to 80°C using a heating rate of 5 °C/min and a frequency of 1 Hz.

Results and Discussion

a. Characterization of the nanosilicas

Figure 1a shows the N_2/77K adsorption isotherms of the nanosilicas with different degree of silanization. The adsorption isotherms belong to the type II of the BDDT classification which corresponds to non-porous materials. The monolayer region is well-defined and the capillary condensation starts at a relative pressure of 0.85. Although it has been claimed that the treatment with silanes does not modify the adsorption capacity of the nanosilicas [3], the H20RD fumed silica (15.5 % silanol) shows less adsorption capacity than the N20 (fully silanated), mainly in the region of low relative pressure (Figure 1b).
The application of the BET equation to the N_2/77K adsorption isotherms shows a reduction in the specific surface area (from 161 m^2/g in the fully silanized nanosilica to 97 m^2/g in the nanosilica containing 15.5 % silanol) and the C_{BET} value (related to the adsorptive-nanosilica interactions) is also reduced from 114 for N20 to 14 for H20RD (Table 2). Therefore, because porosity is not developed in the nanosilicas by treatment with silanes, agglomeration of primary particles in the nanosilica might be produced. The degree of particle agglomeration in the nanosilicas was studied by TEM. Figure 2 shows that when the silanol content decreases, a bigger degree of particles agglomeration is produced.

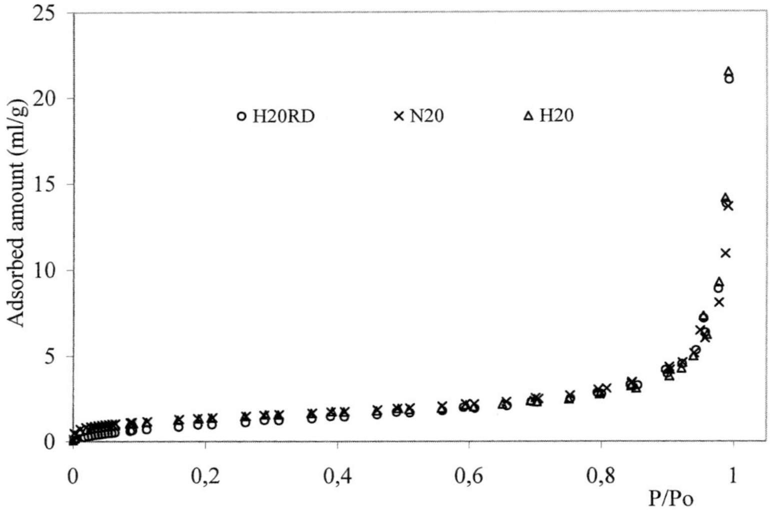

Figure 1a. N_2/77K adsorption isotherms of nanosilicas.

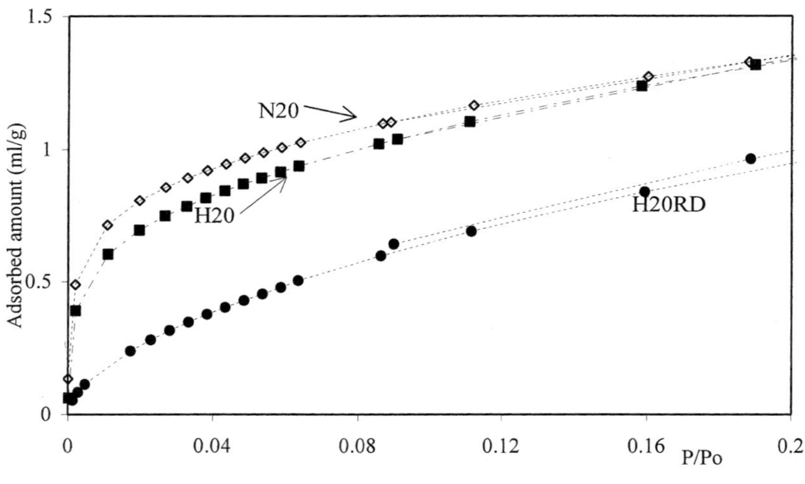

Figure 1b. Low relative pressure region (N_2/77K adsorption isotherms of nanosilicas.

Table 2. Results obtained from the application of the BET equation to N_2/77K adsorption isotherms of nanosilicas.

Nanosilica	SiOH (%)	Specific surface area (m^2/g)		C_{BET} parameter
		Nominal [(a)]	Experimental	
N20	100	200 ± 30	161	114
H20	57.5	170 ± 30	175	55
H20RD	15	200	97	14

a) Data provided by Wacker-Chemie.

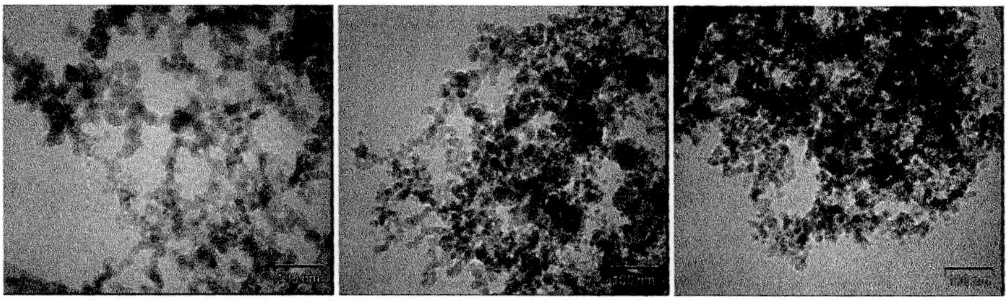

100% silanol 57.5% silanol 15% silanol

Figure 2. Transmission Electron Microscopy – TEM – micrographs of the nanosilicas.

The wettability of the nanosilicas is given in Table 3. The most hydrophilic nanosilica (N20) is well wetted by all polar solvents, whereas the partially silanated nanosilicas are not wetted by water. The higher the carbon content in the nanosilica, the lower the wettability in ethanol.

Table 3. Wettability of nanosilicas.

Nanosilica	Carbon (%)[(a)]	Silanol (%) [(a)]	Water (ml/g)	Water/ethanol (ml/g)	Ethanol (ml/g)
N20	0	100	9.3	11.2	15.8
H20	1.2	57.5	0	15.5	13.7
H20 RD	4.7	15	0	0	9.1

a) Data provided by Wacker-Chemie.

The differences in the wettability in ethanol and water/ethanol mixture of the nanosilicas containing 1.2 and 4.7 % carbon is due to the presence of different amounts of reacted and unreacted silanol surface groups.

The surface chemistry of the nanosilicas was obtained by photoacustic IR spectroscopy (Figure 3). The sharp peak at 3740 cm^{-1} is due to SiOH groups. A very broad peak due to water bonded to the silanol groups on the nanosilica is observed between 3000 and 3600 cm^{-1}. Additionally, bands corresponding to symmetric and asymmetric Si-O-Si absorption are located at 806 and 1110 cm^{-1}, respectively.

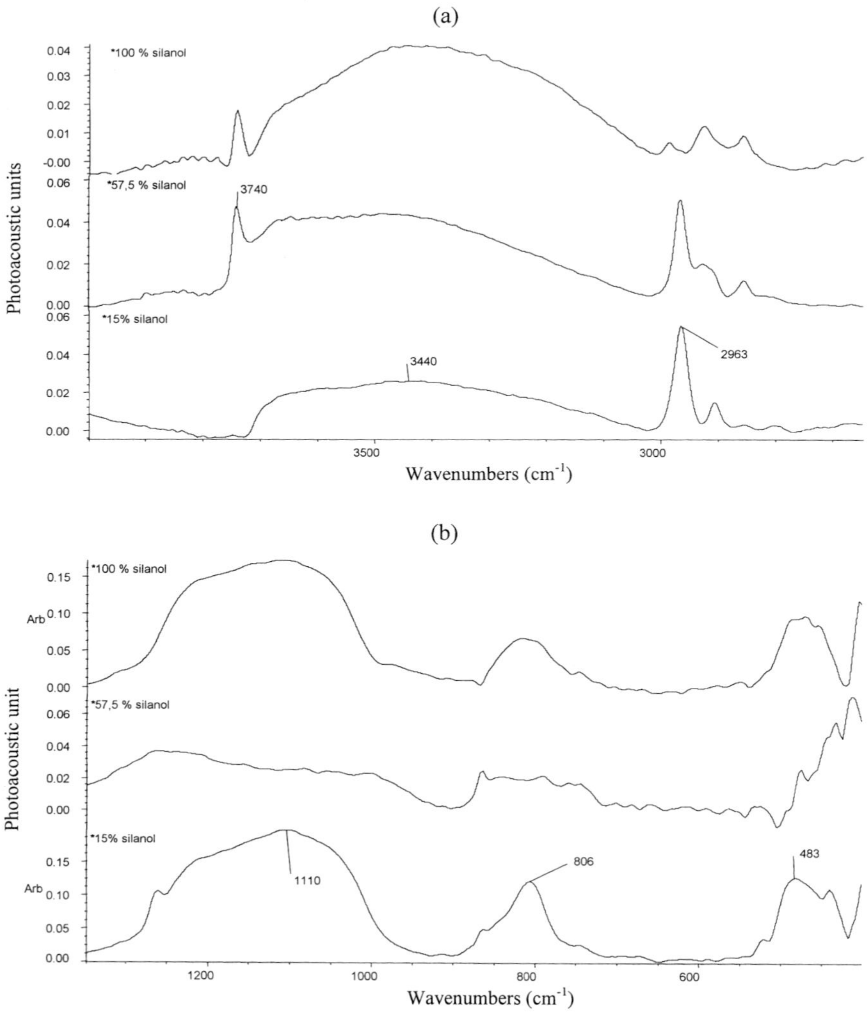

Figure 3. IR spectra of the nanosilicas. (a). 2500-4000 cm^{-1} region. (b) 400-1300 cm^{-1} region.

b. Characterization of the TPU-nanosilica composites

Figure 4a shows the variation of the storage modulus as a function of the temperature for the different nanosilica-TPU composites. The TPU without nanosilica shows lower storage modulus than for the composites containing nanosilicas, and there is a noticeable decrease in the storage modulus by increasing the temperature due to the softening of the polyurethane chains. The higher the silanol content in the nanosilica, the higher the storage modulus and the thermal stability of the nanosilica-TPU composite. This is due to the creation of interactions between the nanosilica and the polyurethane but also the formation of a network of silica particles interacting by hydrogen bonds could contribute.

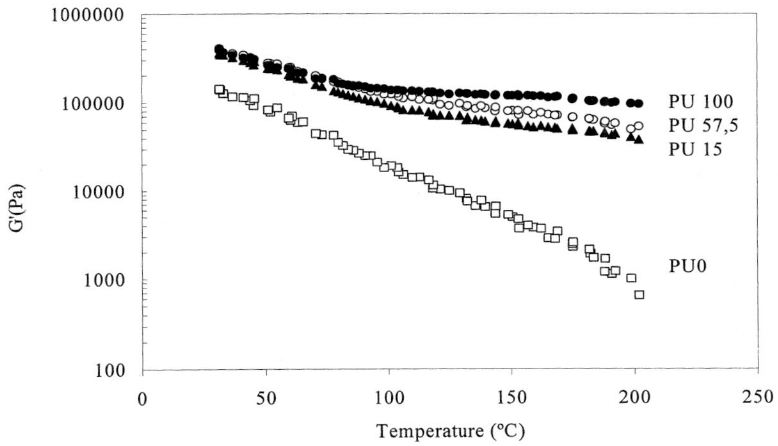

Figure 4a. Variation of the storage modulus as a function of the temperature for the TPU-nanosilica composites.

On the other hand, Figure 4b shows the variation of the storage and loss moduli as a function of the temperature for the nanosilica-TPU composites. The TPU without nanosilica shows a cross over of the loss and storage moduli at 65 °C due to the change in rheological regime of the polyurethane. This cross-over disappears in all TPU-nanosilica composites, and the storage modulus becomes dominant in all of them. The higher the silanol content of the nanosilica, the higher the separation between the storage and loss moduli, i.e. the greater the degree of interaction between the nanosilica and the polyurethane chains. This means that the nanosilica interacts with the TPU chains in the composite, in a greater extent by increasing the silanol content in the nanosilica.

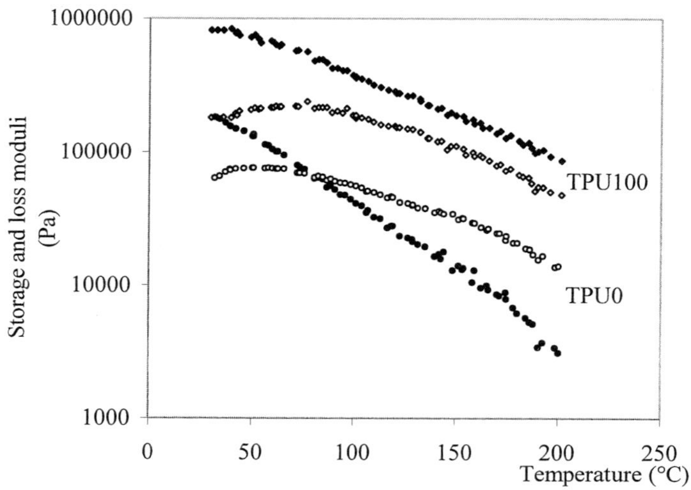

Figure 4b. Variation of the storage (G') and loss (G'') moduli as a function of the temperature for the PU0 and PU100 composites. G', open symbol; G'', closed symbol.

DMTA experiments also confirm the different degree of interactions between the TPU and the nanosilicas with different silanol content. Figure 5 shows the variation of the storage modulus and tan δ vs temperature for the composites. The DMTA curves show the T_g of the soft segments of the polyurethane at about –22°C, and this value does not vary by adding nanosilica. Addition of nanosilica increases the storage modulus and decrease the area under the tan δ curve, as a consequence of a greater degree of crosslinking (physical interactions) and therefore the chain mobility in the polyurethane is reduced by adding nanosilica. The higher the silanol content in the nanosilica, the lower the area under the tan δ curve, which confirms the results discussed earlier. Furthermore, the TPU melts at higher temperature as the silanol content in the nanosilica increases. All those evidences support the relevance of the silanol groups in the interactions of nanosilicas with the polyurethane chains.

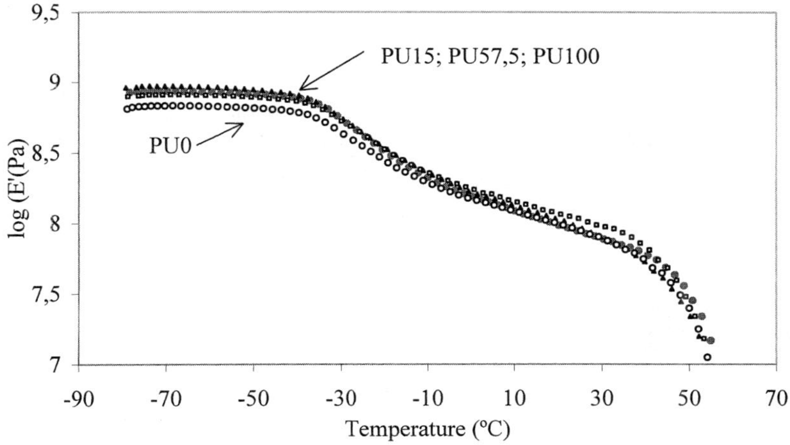

Figure 5a. Variation of the storage modulus as a function of the temperature for the TPU-nanosilica composites (DMTA experiments).

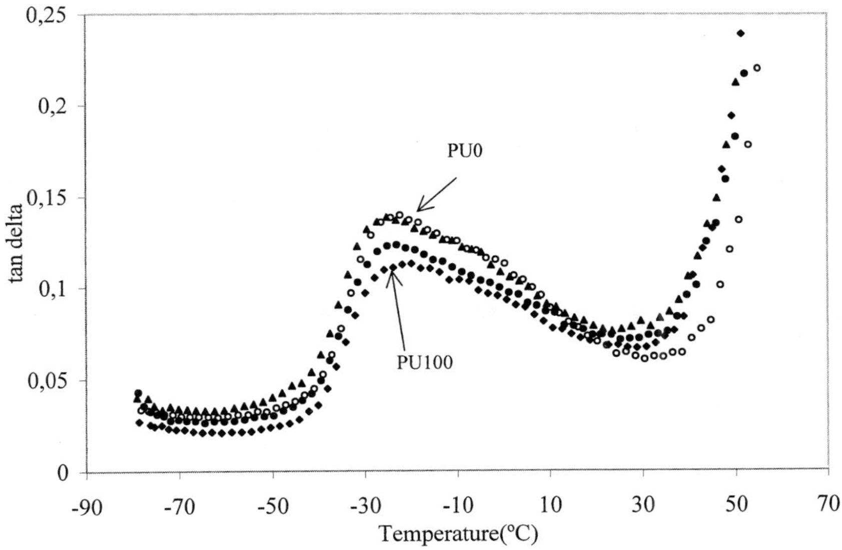

Figure 5b. Variation of the tan δ as a function of the temperature for the TPU-nanosilica composites (DMTA experiments).

Conclusions

Treatment of nanosilicas with silanes increases the degree of agglomeration of the primary particles, giving a reduced specific surface area.

The silanol content in the silica greatly determine the rheological performance of TPU-nanosilica composites, due to the different degree of interactions between the nanosilica and the polyurethane chains. On the other hand, hydrogen bonds between the silanol groups on the nanosilica and the urethane group in the polyurethane are responsible of the improved rheological performance of the nanosilica-polyurethane composites.

Acknowledgments

Authors thank Wacker-Chemie for the preparation of the nanosilicas used in this study, and Dr. A. Hartwig from the Fraunhofer-Institut fuer Fertigungstechnik und Angewandte Materialforschung (Bremen, Germany) for the photoacustic IR measurements of the nanosilicas.

[1] G. Oertel, "*Polyurethane Handbook*", 2rd ed., Hanser, New York, 1993, p.7.

[2] T.G. Maciá-Agulló, J.C. Fernández-García, N. Pastor-Sempere, A.C. Orgilés-Barceló, J.M. Martín-Martínez, *J. Adhesion* **1992**, 38 , 31.

[3] B. Jaúregui-Beloqui, J.C. Fernández-García, A.C. Orgilés-Barceló, M.M. Mahiques-Bujanda, J.M. Martín-Martínez, *J. Adhesion Sci. Tech.* **1999**, 13 , 695.

[4] A. M. Torró- Palau, J. C. Fernández-García, A. C. Orgilés-Barceló, J.M. Martín-Martínez, *Int. J. Adhesion Adhesives* **2001**, 21 , 1.

[5] M.A Pérez-Limiñana, A.M. Torró-Palau, A.C. Orgilés-Barceló, J.M. Martín-Martínez, *Macrom. Symp.* **2001**, 169, 191.

Complex Adhesion Effects of Inorganic Nanofillers *vs* Microfillers in Polymer Composites

V. Kovačević, M. Leskovac, S. Lučić Blagojević, D. Vrsaljko*

University of Zagreb, Faculty of Chemical Engineering and Technology, Marulićev trg 19, 10000 Zagreb, Croatia
Fax: + 385 1 4597 260; E-mail: vkovac@fkit.hr

Summary: The adhesion parameters of fillers in composites, coefficient of wetting, interfacial free energy and work of adhesion, were calculated in order to evaluate the difference between the effects of selected microfillers *vs* nanofillers in polyacrylate (PA) and poly (vinyl acetate)(PVAc) matrix. The results showed that the improved fundamental adhesion, i.e. higher work of adhesion in nanocomposites resulted in the improved practical adhesion i.e. higher mechanical properties. The interfacial region was additionally modified by the controlled pre-treatment of calcium carbonate micro- and nanofillers by sodium stearate that lowered the interactions with PVAc matrix. The composite morphological and mechanical properties changed accordingly. The lowered coefficient of interactions explained the pronounced composite failure at the interphase by dewetting and finally lead to the composite weakening. The conclusion about the complex relationship between the composite properties and engineering the interphase was confirmed.

Keywords: adhesion; failure, micro and nanofillers; nanocomposites; surface parameters

Introduction

Adhesion science might be applied in investigation of composites, as the materials where the three-dimensional area between matrix and filler, i.e. interphase, has a crucial role in determination of composite properties.[1] The interest in adhesion science is a result of its applicability to solution the engineering problems such as bounded components in adhesive joint but also in composite materials.[2] Of the key importance in adhesion is generally wetting and setting.[3] We found that the adhesion parameters in particulate filled composites, such as coefficient of wetting (S), interfacial free energy ($\gamma_{\text{filler/matrix}}$) and work of adhesion ($W_{\text{filler/matrix}}$) (eqs.1-3), could explain the composite behaviour.[4]

 DOI: 10.1002/masy.200550302

$$S_{\text{filler/matrix}} = \gamma_{\text{filler}} - \gamma_{\text{matrix}} - \gamma_{\text{filler/matrix}} \quad (1)$$

$$\gamma_{\text{filler/matrix}} = \gamma_{\text{filler}} + \gamma_{\text{matrix}} - 2(\gamma_{\text{filler}}\, \gamma_{\text{matrix}})^{1/2} \quad (2)$$

$$W_{\text{filler/matrix}} = \gamma_{\text{filler}} + \gamma_{\text{matrix}} - \gamma_{\text{filler/matrix}} \quad (3)$$

In equation (2) the Good - Girifalco interaction parameter is taken as unity, when the individual polar and dispersion components of the composite components appear not to be available.[5]

An increasing understanding of adhesion science enables engineers to develop the ways in which interfacial region can be modified.[2] By modifying the interface and by controlling the adhesion in composites, adhesive joints and/or multi-layers the resulting properties could be changed accordingly.[3]

The characteristics of filler particles that have the significant influence on the composite properties are shape and size, specific surface area, packing and surface activity, distribution and interactions with matrix.[6] On the other hand the polymer matrix characteristics that are relevant are flexibility/rigidity of polymer chains, crystallisation ability and also surface activity.[7]

The difference between the effects of nanofillers in nanocomposites *vs.* microfillers in the conventional microcomposites exists in the dimension of dispersion of the two components, organic polymer matrix and inorganic filler.

In the nanocomposite a great amount of matrix is affected by ultra-fine phase of nanofillers, that create the special 'constrained' matrix structure[8] with the restricted mobility of polymer chains.[9]

The modification of matrix/filler adhesion, by lowering the filler size to nano-dimensions *vs.* micro-dimensions and/or by changing the filler surface characteristics by the controlled filler surface pre-treatment, has been the subject of investigations in this paper.

The aim is to investigate the change in adhesion at the interphase and the resulted morphological and mechanical properties in poly(vinyl acetate) (PVAc) and polyacrylate (PA) composites, as a consequence of the effects of nanofillers *vs.* microfillers and the selected pre-treatment of the nanofiller surface.

Experimental

The characteristics of matrices and fillers presented in Table 1 are determined previously.[4,10-12] The general aim was to compare the effects of kaolin filler with the great surface area due to layered structure [12] and thus well suited to create nanocomposites *vs.* diatom microfiller in the same PA matrix. The other example presented in Table 1 is nano-sized (<100nm) calcium carbonate ($CaCO_3$) filler *vs.* microsized $CaCO_3$ in PVAc matrix. The investigations are made on the thin films of composites (thickness ≈0.2 mm), prepared by pouring the composition on polyethylene foil and drying to the constant weight. The composite samples were prepared by simple mixing with necessary additives and different filler volume fraction (ϕ_f = 3,6,9,12 %). The details of the preparation of microcomposites and nanocomposites used in this paper are described elsewhere.[10,11]

Table 1. Structure and properties of matrices and nanofillers (N) *vs.* microfillers (M).

Sample	Characteristics			
Matrix	PA		PVAc	
1) Structure	*		**	
2) T_g(°C)	20.4		18.1	
3) γ (mJm^{-2})	54.8		37	
4) Filler	Kaolin (N)	Diatom (M)	$CaCO_3$ (N)	$CaCO_3$ (M)
3) γ (mJm^{-2})	72.4	60.5	72.9	58
5) x_{50} (nm)	600	1300	80	520
6) S_{BET}(m^2g^{-1})	11.63	2.61	12	0.96

1)* Polyacrylate (PA): Styrene-co-butylacrylate-ethylbenzene-acrylamide-formaldehide water dispersion (50 wt%) (BASF, Germany).
** Poly(vinyl acetate)(PVAc): water emulsion (55wt%) (Karbon, Croatia).
2) DSC method [11,12].
3) Surface energy measured by methods: contact angle, IGC, capillary [4, 10].
4) $CaCO_3$ (M): Calcite (Industrochem, Croatia);$CaCO_3$ (N): Precipitated $CaCO_3$ (Solvay, Germany); Diatom Opal C [11], (Solvay, Germany); Kaolin China clay [12] (ECC International, UK).
5) Amount of 50% particles with sizes lower than the value given (suppliers data).
6) Specific surface area [11, 12].

A JEOL JSM-330 scanning electron microscope was used for investigation the composite surface structure. In order to evaluate the initial morphology of nanocomposites *vs.*

microcomposites and the changes in morphology after failure, the composite films were observed before and after fracture was produced.

Tensile properties of the composite films were determined with a Zwick 1445 Universal testing Machine, with a crosshead speed of 50 mm/min and 50 mm gauge length.

The controlled surface pre-treatment of nanofiller calcium carbonate samples by adding the increasing amount of sodium stearate to change the filler surface activity, was carried out. This is described in a previous paper.[10]

Results and Discussion

The relevant factors for the composite study are the adhesion phenomena at the interphase (Table 2).

Table 2. Adhesion effects of nanofillers (N) *vs.* microfillers (M) at the interphase in composites.

Samples		Interphase in composites (mJm^{-2})		
Matrix	Filler	[1] S_{fm}	[2] γ_{fm}	[3] W_{fm}
PA	Kaolin (N)	16.4	1.2	139.0
	Diatom (M)	-5.3	0.2	118.5
PVAc	$CaCO_3$ (N)	29.9	6.06	104.0
	$CaCO_3$ (M)	18.6	2.36	92.4

[1] Equation (1); [2] Equation (2); [3] Equation (3).

Results of studies on the effective adhesion for a given system indicate some conditions as optimal, i.e. work of adhesion as maximal (optimal), interfacial free energy as minimal (tends to null) and coefficient of wetting as a positive value.[3]

It should be stressed that the improved fundamental adhesion might result in the improved practical adhesion.[3] For example, the work of adhesion is a term, which has a strong, but indirect influence on practical adhesion. We found that although the work of adhesion levels are low (of order of mJm^{-2}) compared with fracture energies (often several kJm^{-2}), that the small absolute increases in the work of adhesion lead to large absolute increases in fracture energy.[10] The increase in interfacial strength between filler and matrix and thus in work of adhesion (W_{fm}) enables more bulk energy dissipation to occur, for example through plastic energy losses.[13] The results in Table 2 show that a higher value of work of adhesion would be achieved with the nanofillers. Of the great importance for the improvement of the composite properties is a good filler dispersion in the polymer matrix, which depend on the process of composite preparation, wetting ability and strength of interactions between filler and matrix.

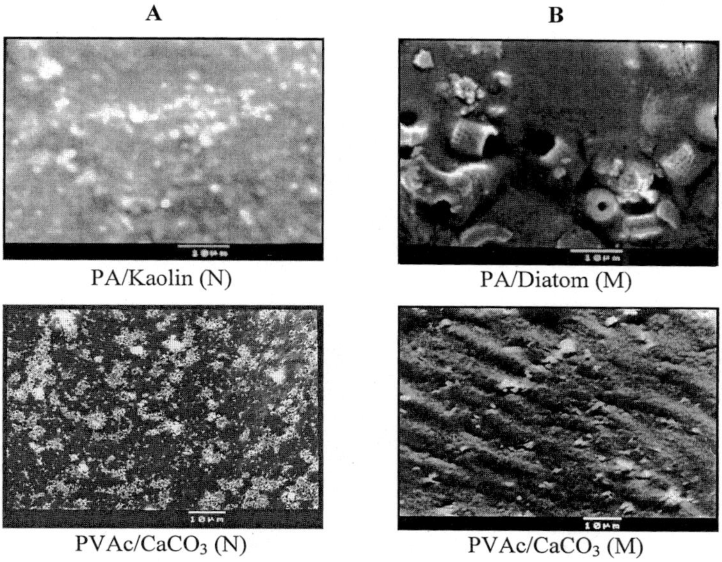

Fig. 1. Morphology of composites filled with nanofillers (A) and microfillers (B) ($\phi_f = 0.12$).

The morphologies of nanocomposites *vs.* microcomposites are illustrated by the 'net-like' structure of nanofiller particles (Figure 1A), in comparison to the micro-particles and their agglomerates dispersed as isolated 'islands' in the matrix (Figure 1B). The specific structure of nanocomposites with significant volume of organic phase affected by nanofiller, is described in literature as similarity with ionomers structure, where the strong interactions determined the phase heterogeneity and specific properties of the system.[9]
The polymer matrix introduction into the nanofiller three-dimensional 'net-like' structure results in a matrix 'constrained' structure.[8,14] The kaolin filler, which is supposed as a

nanofiller due to its layered structure and high specific surface area (Table 1),[12] showed the same "net-like" effect on the composite morphology as we noticed with $CaCO_3$ nano-sized filler particles (Figure 1A). The similar effects on composite morphology were illustrated in literature for poly (methyl methacrylate) composites filled with the same volume fraction of aluminium hydroxide of coarse and/or fine filler particles.[7]

The strong interactions are expected to change the locus of failure. In adhesion study, the locus of failure, or separation between phases is also very important, i.e. cohesive in matrix *vs.* adhesive failure at the interphase.[12]

In the case of evenly distributed filler particles the high specific surface area of nanofillers could assure the increased amount of the interphase between matrix and filler and thus the significant changes in the nanocomposite properties. It is necessary to stress the importance of the successful dispersion of nanofillers in the matrix.

The interphase has a crucial role as a 'heart' in composites[15] due to its importance in transferring the stresses.

Failure in the material is taking place at the weak points in structure, where the product EG, in equation (4), i.e. composite modulus times fracture energy in the known Griffith-Irwing theory of fracture, has the lowest value.[6]

$$\sigma_F = k\,(EG / l\,)^{1/2} \tag{4}$$

where σ_F is a fracture stress, k is a constant and l is the length of critical crack.

Mechanisms of failure in composites could take place in polymer matrix by shear yielding and/or crazing, inside the aggregates of filler particles and/or at the interface that depending on the 'weak' points where the product EG has the lowest value. In Figure 2B it is visible that in microcomposites the failure is taking place at the interphase by dewetting. On the other hand in the case of small radius of particles (R) of nanofillers, the stress at dewetting, σ^D, is enough high to transfer the failure from the interphase to the matrix (Figure 2A), because of the high work of adhesion between filler and matrix, W_{fm} in equation (5):

$$\sigma^D = - C_1\, \sigma^T + (C_2\, W_{fm} / R)^{1/2} \tag{5}$$

σ^T is the thermal stress that, in our isothermally prepared composites, should be close to zero. *C1* and *C2* are constants.

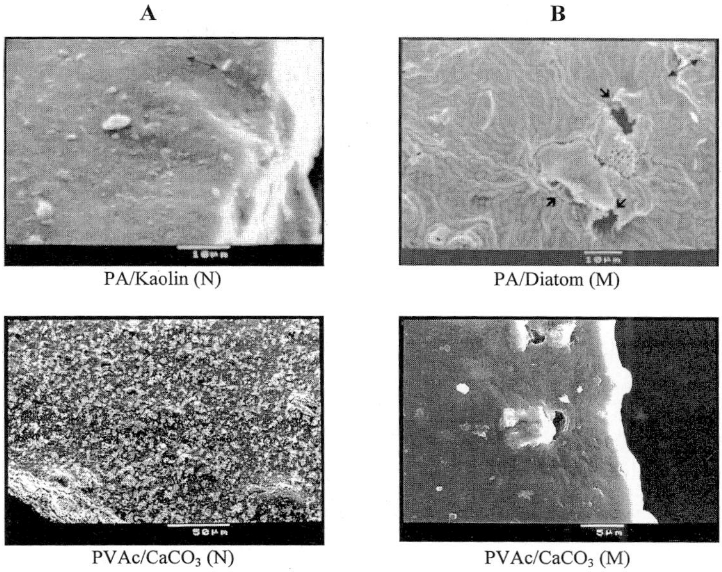

Fig. 2. Failure in nanocomposites (A) and microcomposites (B) (ϕ_f= 0.12).

The examples from literature illustrate cohesive failure in matrix due to strong interactions and dewetting at the interphase due to low interactions.[16] The conclusion could be drawn that the stress need to initiate the failure in composite increased with decreasing radius of filler particles. In that case the interphase is not any more the weak point in a composite.

We found the improved fundamental adhesion resulted in the improved practical adhesion.[12] The higher work of adhesion and improved wetting in nanocomposites *vs.* microcomposites in Table 2 with no signs of dewetting (Figure 2), should thus result in the improved mechanical properties i.e. composite reinforcing (Figure 3).

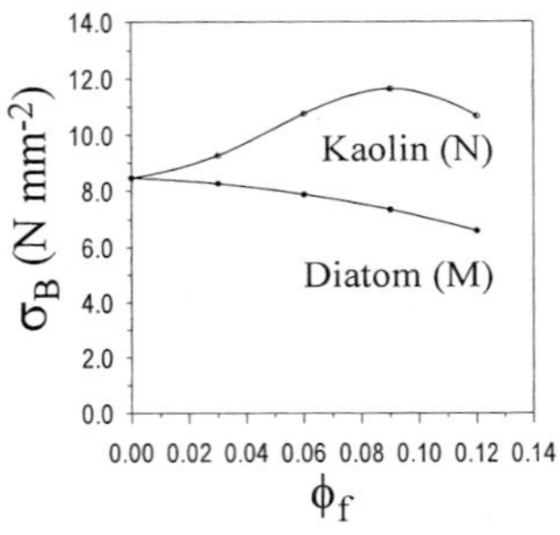

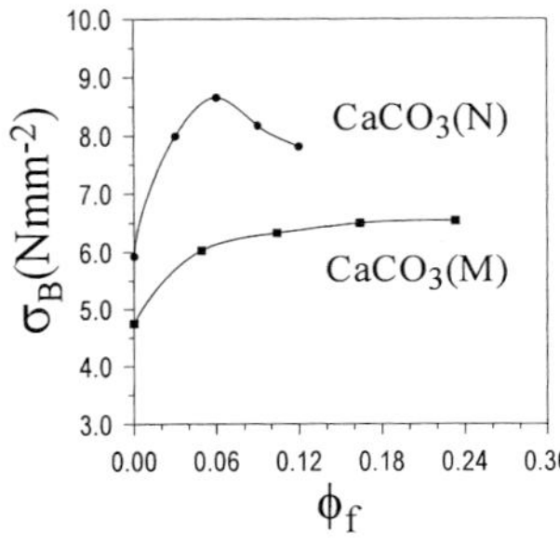

Fig. 3. Effect of fillers on the composite strength at break (σ_B) with increased filler vol. fraction (ϕ_f); N= nanoparticles, M= microparticles.

The results in Table 2 showed that the adhesion parameters at the interphase changed as a consequence of the effects of nanofillers *vs.* microfillers.

After the $CaCO_3$ nanofiller pre-treatment, the results in Table 3, i.e. the negative wettability, the increased interfacial free energy along with the lowering in the work of adhesion illustrate that a relationship exists between the thermodynamic parameters, which characterize the polymer-filler interactions at the interphase.

Table 3. Interfacial adhesion parameters before and after the $CaCO_3$ filler pre-treatment with sodium stearate.

Composite	[1] Interphase (mJm^{-2})		
	S_{fm}	γ_{fm}	W_{fm}
PVAc/$CaCO_3$ (M)	18.6	2.36	92.4
[2] PVAc/$CaCO_3$ (MS)	17.2	2.00	91.2
[3] PVAc/$CaCO_3$ (N)	29.86	6.06	103.9
PVAc/$CaCO_3$ (NS_1)	22.09	3.29	96.1
PVAc/$CaCO_3$ (NS_2)	-9.64	0.63	64.4
PVAc/$CaCO_3$ (NS_3)	-18.28	2.26	55.7
PVAc/$CaCO_3$ (NS_4)	-25.30	4.33	48.7

[1] Table 2; [2] Commercial treatment; [3] Filler surface coverage [10] : N (0%); NS_1 (16%); NS_2(47%); NS_3(78%); NS_4(< 100%) calculated using 0.21 nm^2 as the 'cross-sectional area' for stearate.[17]

The changes of the adhesion parameters in composites by the interface engineering as a consequence of the controlled filler surface pre-treatment (Table 3) are expected to reflect in the changes of the composite properties (Figure 4).

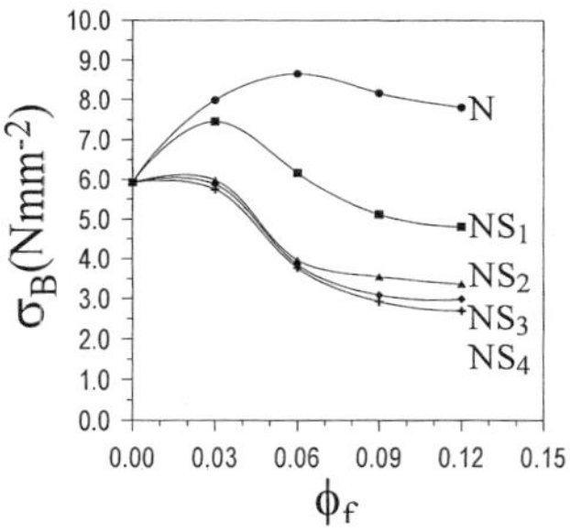

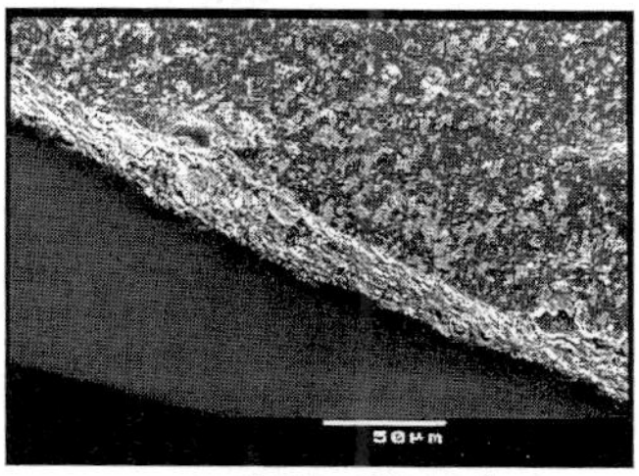

$CaCO_3$ (N)

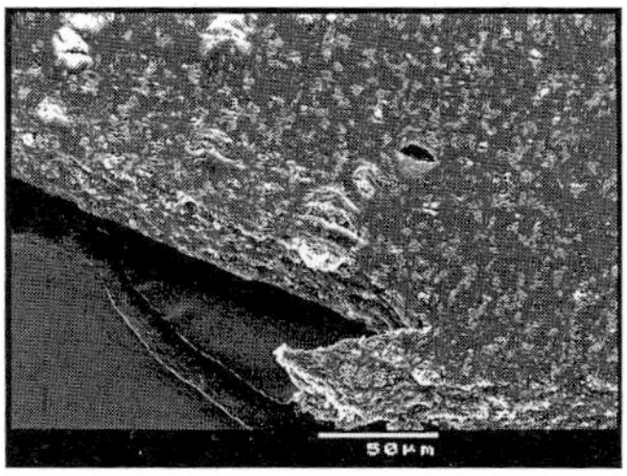

$CaCO_3$ (NS_1)

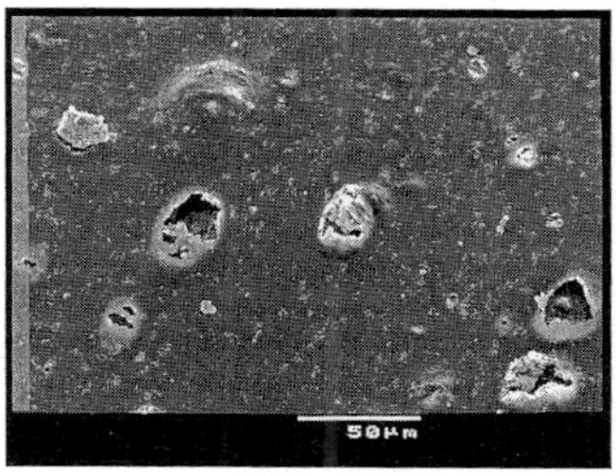

$CaCO_3$ (NS_2)

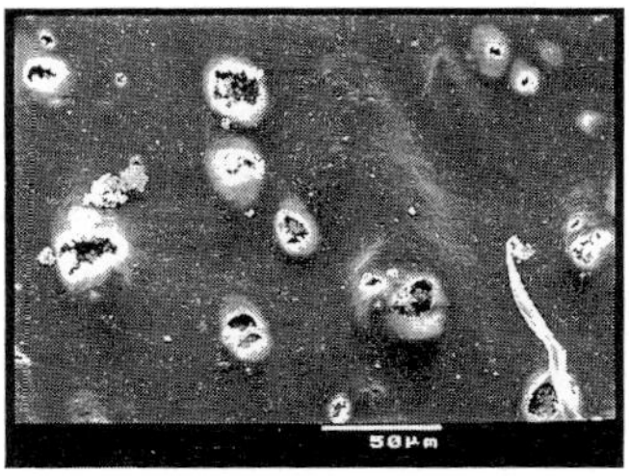

$CaCO_3$ (NS_3)

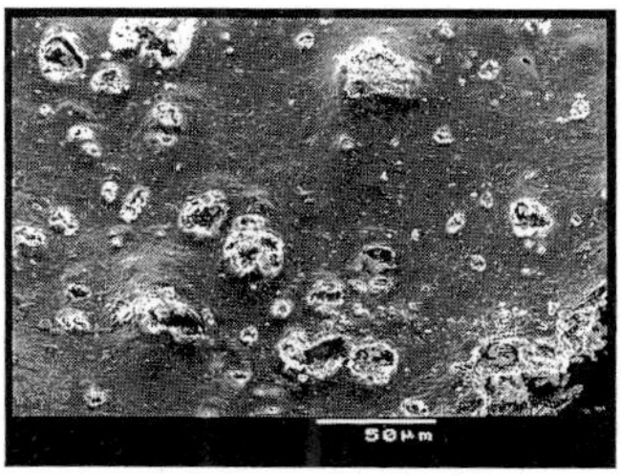

$CaCO_3$ (NS_4)

Fig. 4. Effect of the $CaCO_3$ nanofiller (N) increased pre-treatment (NS_1-NS_4) on the PVAc composite strength at break and the resulted morphology at failure.

The reinforcing effect in the composite PVAc/$CaCO_3$ (N) with the excellently adhering nanoparticles is followed by the high work of adhesion (Table 3) and thus no signs of dewetting at the interphase, while the failure is directed into the polymer where dissipation of fracture energy (high local G in equation 4) leads to a higher fracture stress. On the other hand the opposite effects are found after the filler increased pre-treatment (NS_1 - NS_4) (Table 3 and Figure 4).

One can conclude that the relationship between the properties of composites and engineering the interfacial region by pre-treatment is very complex in a way that often the modification could improve one property but will have an adverse effect on another.

The quantitative calculations of the interactions by fitting the mechanical properties of filled composites to the theoretical and/or empirical equations might help in the prediction of the composite mechanical properties. The relationship between the ultimate strength and relative amount of particulate filler described by the exponential model proposed by Pukanszky[18] gives the interaction coefficient B:

$$B = (1 + LA_f \rho_f) \ln \sigma_{Ti} / \sigma_{To} \tag{6}$$

where A_f and ρ_f = specific surface area and density of filler; L and σ_{Ti} = thickness and strength of the interphase; σ_{To} = true tensile strength of the polymer matrix.

Table 4. Comparison of the interactions between matrix and fillers.

Samples	[1)] B
PA/Diatom (M)	3.8
PA/Kaolin (N)	8.6
PVAc/$CaCO_3$ (M)	6.4
PVAc/$CaCO_3$ (MS)	5.3
PVAc/$CaCO_3$ (N)	9.9
PVAc/$CaCO_3$ (NS_1)	5.9
PVAc/$CaCO_3$ (NS_2)	2.3
PVAc/$CaCO_3$ (NS_3)	1.4
PVAc/$CaCO_3$ (NS_4)	0.9

[1)] Equation (6).

The quantification of interactions between filler and matrix at the interphase, helps in the evaluation and comparison of the composite mechanical and other properties (Table 4). It is important to stress that the interactions coefficients should be used only for comparison purposes. Still we showed that the parameters from equations used for fitting the experimental data to various models are strong indicators of polymer-filler adhesion.[12] For example, the higher interaction coefficient in Table 4 means the higher work of adhesion at the interphase of composite (Table 3). On the other hand, the increased filler surface pre-treatment (NS_1 – NS_4) lowered the interactions and thus the coefficient *B* resulting in the composite weakening (Figure 4).

Conclusion

The possibilities of 'engineering the interphase' in composite by lowering the filler size to nanodimensions and increasing the active surface area or changing the adhesion parameters by surface pre-treatment, may alter the composite properties in the interfacial region and thus in the composite material as a whole.

Acknowledgements

We are grateful for support from the Ministry of Science and Technology of Croatia and Universities of Bath and Loughborough (UK) for facilitating the international collaboration.

Thanks are due to Dr. Ch. Nover (Solvay Alkali GmbH, Germany), for supporting us with samples and information, and Dr. D.E. Packham for advices given during the experiments.

[1] P.E. Packham, in: *Handbook of Adhesion*, D.E. Packham (Ed.), Longman Scientific & Technical, Harlow, Essex, **1992**, pp.78-79.
[2] D.E. Packham, *National Meeting of the American Institute of Chemical Engineers*, Proceedings of the Conference, Seattle, USA, August **1993**, 24a.
[3] K.L. Mittal, *Pure and Appl. Chem.* **1980**, 52, 1295-1305.
[4] V. Kovačević, M. Leskovac, S. Lučić Blagojević, *J. Adhesion Sci. Technol.*, **2002**, 16(14), 1915-1929.
[5] A.J. Kinloch, *Adhesion and Adhesives, Science and Technology*, Chapman & Hall, London, **1995**.
[6] D.E. Packham, *J. Adhesion*, **1995**, 54, 133-143.
[7] R. Rothon in: *Particulate-Filled Polymer Composites*, Longman, Harlow, UK, 1995, pp.31-39.
[8] H.L. Frisch, J.E. Mark, *Chem.Mater.*, **1996**, 8, 1735-1738.
[9] G. Tsagaropoulos, A. Eisenberg, *Macromolecules*, **1995**, 28, 396-398.
[10] S.Lučić, V. Kovačević, D.E. Packham, A. Bogner, A. Geržina in: *Polymer Surface Modification: Relevance to Adhesion*, K.L. Mittal (Ed.), VSP **2000**, Vol. 2, p.505-525.
[11] M. Leskovac, V. Kovačević, S. Lučić, H.R. Perrott, I. Šmit, *Mat.Res.Innov.*, **2002**, 6, 206-213.
[12] V. Kovačević, S.Lučić Blagojević, M. Leskovac, *J. Adhesion Sci. Technol.*, **2002**, 16(10), 1343 – 1365.
[13] D.E. Packham. in: *Mittal Festschrift on Adhesion Science and Technology*, W.J. van Ooij and H.R. Anderson, Jr.(Eds), VSP, Utrecht, 1998, p.81.
[14] T. Seagusa, *Macromol. Symp.* **1995**, 98, 719-729.
[15] M.R. Piggot, A. Sandi, P.S. Chua, D. Anderson, in: *Composite Interfaces*, H. Ishida & J.L. Koenig, North Holland, New York, **1996**.
[16] M.E. Dekkers, D. Heikens, *J. Materials. Sci.*, **1984**, 19, 3271-3275.
[17] E. Papirer, J. Schultz and C. Turchi, *Eur. Polym. J.* **1984**, 20, 1155-1158.
[18] B. Pukanszky, *Composites*, **1990**, 21, 255-262.

Influence of the Particle Size of $CaCO_3$ on the Adhesion of Filled EVA Materials

Asunción Martínez-García,[1] Ana Sánchez-Reche,[1] Carmen M. Cepeda-Jiménez,[2] José Miguel Martín-Martínez[2]*

[1] AIJU. Toy Research Institute, Avda. de la industria, 23, 03440 Ibi (Alicante), Spain
E-mail: proyectos@aiju.info
[2] Adhesion and Adhesives Laboratory, Department of Inorganic Chemistry, University of Alicante, 03080 Alicante, Spain
E-mail: jm.martin@ua.es

Summary: A surface treatment with corona discharge was used to improve the adhesion properties of ethylene vinyl acetate copolymer (EVA) containing small amounts of four $CaCO_3$ with different particle size. The nature of the surface modifications produced by the corona discharge treatment and the adhesion to a polychloroprene adhesive were assessed. Treatment of $CaCO_3$ filled EVA with corona discharge produced a decrease in water contact angle value, irrelevant to the different particle size of the calcium carbonates. The corona discharge treatment created C-O and C=O moieties on the EVA surface and also increased the peel strength, more markedly as the $CaCO_3$ particle size increased. In general, a mixed (adhesion + cohesive in the EVA) failure in the filled EVA material was produced (assessed by IR-ATR spectroscopy and SEM micrographs of the failed surfaces), but the failure was more cohesive in the EVA containing higher particle size $CaCO_3$. The durability of the joints was also studied.

Keywords: adhesion; $CaCO_3$; corona discharge; EVA; IR spectroscopy; T-peel strength

Introduction

The particle size of the fillers usually determines the filler-polymer interactions and the mechanical properties of polymers [1]. In general, the smaller the filler particle size, the better the mechanical properties, but the viscosity is poorer and the mixture is more difficult to process [2], due to the agglomeration of primary particles. As the particle size of $CaCO_3$ increases, its specific surface area decreases. Adhesion performance of polymers is also affected by the filler nature and particle size. Therefore, in this study, corona discharge was used as a surface treatment to improve the adhesion properties of two EVA copolymers containing four $CaCO_3$ with different particle size. Corona discharge is a fast treatment, it can be easily adapted to industrial production, and it produces an increase in the adhesion of several polyolefins [3-5].

 DOI: 10.1002/masy.200550303

Experimental

The EVA12 pellets were supplied by Repsol Química, S.A. (Santander, Spain). The melt flow index of EVA12 is 0.6 g/10 min and its Shore A and D hardnesses are 96 and 45, respectively. Unfilled and filled EVA12 were moulded using a Margarit JSW injection machine, and the following conditions were used : Temperature = 170 °C; injection time = 6 s; injection pressure = 60 % of total pressure (1570 bar; 1 bar=10^5 Pa); mould temperature = 20 °C. The copolymer was filled during the injection process with 5 wt% of four ultramicronized and calcium stearate surface treated $CaCO_3$ fillers (supplied by S.A. Reverté, Bellvei, Spain). The four $CaCO_3$ have different mean particle sizes (0.6, 1.0, 1.8 and 4.0 μm) and different specific surface areas (14.5, 11, 6 and 3 m^2/g, respectively), and in this study they are referenced as T1, T2, T3 and T4, respectively. The injection moulded pieces were cut into test samples of sizes 20×30×2 mm for characterization and 150×30×2 mm for adhesion tests.

Adhesive joints of treated $CaCO_3$ filled EVA12 were made using a commercial two-component solvent-based polychloroprene adhesive consisting of polychloroprene adhesive itself (Telcopren 3003, supplied by Composan Adhesivos S.A., San Vicente del Raspeig, Spain) and 5 wt% polyisocyanate (Desmodur RF, provided by Bayer, Leverkusen, Germany). The adhesive and the isocyanate were mixed just before application using a brush on two identically treated filled EVA12 strips. The adhesive was cured in open air for 1 hour and afterwards melted at 100°C by using an infrared lamp (to facilitate the interlocking of the chains of the adhesive). The two strips were then placed in contact and a pressure of 0.8 mPa was applied for 10 seconds to achieve a suitable joint. The polychloroprene adhesive solution contains 25 wt% solids and its Brookfield viscosity is 3.0 ± 0.2 Pa.s. The thickness of the adhesive layer applied on EVA12 was about 100 μm.

A Tantec Corona Generator model HV09 was used to modify the unfilled and $CaCO_3$ filled EVA. The distance between the upper electrode and the sample was adjusted to 1 mm. The sample was located on the counterpart nylon electrode, which could be displaced at a controlled speed. A hook-shaped electrode was used, and the treatment was performed in air at atmospheric pressure. The discharge power was set to 93 W and 150 mm long EVA samples were passed under the electrode at a speed of the counterpart nylon electrode of 19 mm/s. These conditions provide a corona energy of 3.3 J/cm^2. The corona energy was obtained using the following equation:

$$\text{Corona energy} = \frac{\text{Power}}{\text{Speed x treatment length.}}$$

Contact angle values on the treated $CaCO_3$ filled EVA12 were measured at 25 °C using a Ramé-Hart 100-0 equipment by placing 4 μl drops of deionized and doubly distilled water on the surface. The samples were placed into a chamber saturated with water vapour at 25 °C, at least 15 min before the drop was placed on the filled EVA12 surface. The contact angle values were measured immediately after corona discharge treatment. The experimental error was ± 2 degrees.
ATR-IR spectra of corona discharge-treated $CaCO_3$ filled EVA12, before and after T-peel tests, were obtained using a Nicolet FTIR 550 spectrometer. To avoid deep penetration of the IR radiation into the sample, the attenuated total multiple reflection method was employed using a KRS-5 (thallium bromo-iodide) crystal. The incident angle of the IR radiation was 45°. 100 scans were averaged at a resolution of 4 cm^{-1}.
SEM micrographs of the as-received and corona discharge treated unfilled and $CaCO_3$ filled EVA12, before and after T-peel tests, were obtained in a JEOL SEM J840 instrument using an electron beam energy of 20 kV. The samples were gold coated and the electron beam was 20 kV.
The adhesion was evaluated from T-peel tests on as-received and treated $CaCO_3$ filled EVA12/polychloroprene adhesive joints (72 hours after bond formation) using an Instron 4011 instrument; a crosshead speed of 0.1 m/min was used. Five replicates for each experimental variable were obtained and data were averaged with an error less than ± 0.5 kN/m.

Results and Discussion

Figure 1 shows the SEM photographs of the four $CaCO_3$ where the shape and the different degree of agregation of the particles can be appreciate. As the $CaCO_3$ particle size decreases, the degree of agregation is higher.
The ATR-IR spectra of the 5 wt% $CaCO_3$ filled EVA12 (Figure 2) correspond to that of the EVA12 and also typical bands of the calcium carbonate at 1437 and 874 cm^{-1} appear, indicating that a physical mixture of the components was obtained.

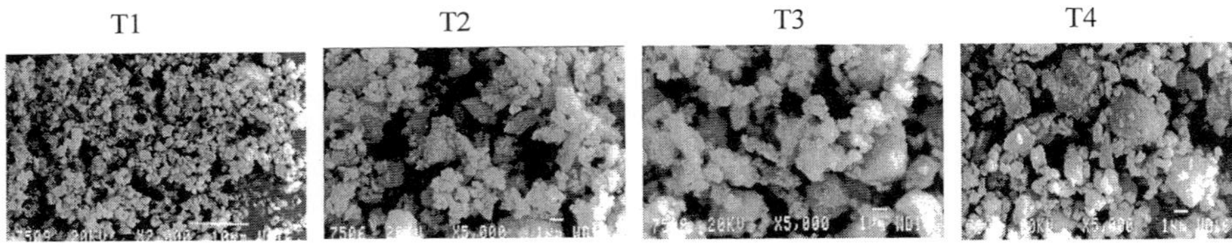

Figure 1. SEM micrographs of the $CaCO_3$ fillers.

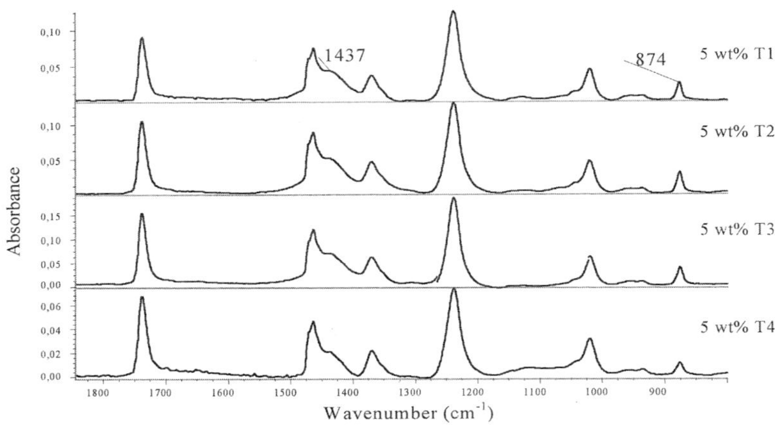

Figure 2. ATR-IR spectra of the different particle size $CaCO_3$ filled EVA12.

The corona discharge treatment of the different size $CaCO_3$ filled EVA12 was carried out at a corona energy of 3.3 J/cm^2. The contact angle values of filled EVA12 (Figure 3) decrease from 89-94 degrees (untreated) to 53-48 degrees (after corona discharge treatment), and these values are similar for all EVA12 containing $CaCO_3$ of different size.

The ATR-IR spectra of the corona discharge-treated T1, T2 and T3 $CaCO_3$ filled EVA12 show that the carbonate bands at 1437 and 874 cm^{-1} are more intense than that of the untreated material, indicating that the treatment exposes $CaCO_3$ particles to the surface. The ratio of the relative intensities at 874 cm^{-1} (typical of $CaCO_3$) and 1238 cm^{-1} (typical of EVA12) bands increases after corona discharge treatment (Table 1); that is, some EVA is removed by treatment with corona discharge and some $CaCO_3$ particles become exposed to the surface.

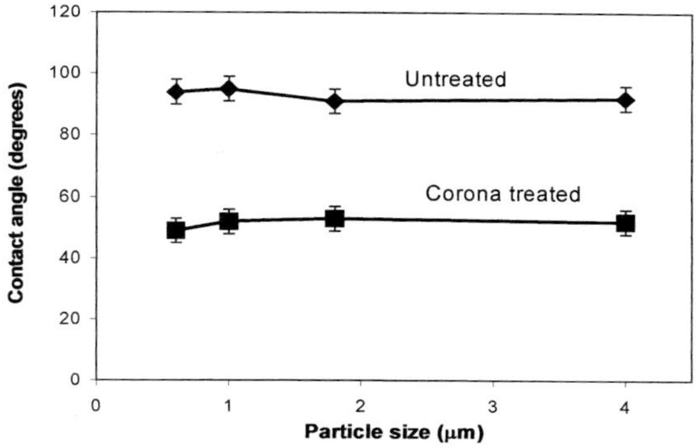

Figure 3. Contact angle values (water, 25°C) on EVA12 filled with 5 wt% $CaCO_3$ of different particle size.

Table 1. Ratio of the relative intensities at 874 cm^{-1} ($CaCO_3$) and 1238 cm^{-1} (EVA12) bands (ATR-IR spectra) in the EVA12 filled with 5 wt% $CaCO_3$ of different particle size.

Material	Untreated	Corona discharge-treated
EVA-T1	0.21	0.50
EVA-T2	0.24	0.44
EVA-T3	0.24	0.46
EVA-T4	0.15	0.17

The ratio of the relative intensities in the corona discharge-treated T4 filled EVA12 is similar to that of the untreated material, which could indicate that the filler particle size affects the interaction of $CaCO_3$ with EVA12 and also the extent of the modifications produced by the corona discharge treatment. By increasing the particle size of the $CaCO_3$ a lower number of primary particles tend to agglomerate and this will allow a more homogeneous mixture with EVA12 during processing, in agreement with previous studies [6].

On the other hand, bands at 1460-1500 cm^{-1} and 1640 cm^{-1} appear in all ATR-IR spectra of the corona-discharge treated materials (Figure 4) indicating the formation of carbon-oxygen moieties (mainly C-O and C=O) [5].

As a consequence of the chemical and morphological modifications produced by corona discharge treatment, improved adhesion of $CaCO_3$ filled EVA12 is achieved (Table 2). Similar adhesion was obtained in the untreated materials filled with T2, T3 and T4 (0.7 kN/m). These values increase after corona discharge treatment, more markedly for the joint produced with the EVA filled with the higher particle size $CaCO_3$ (T4).

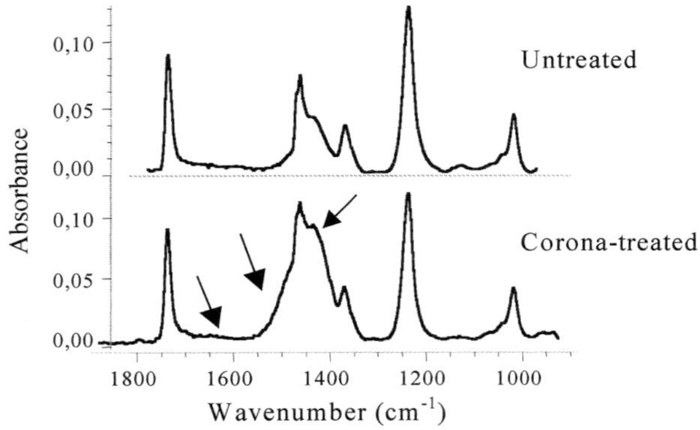

Figure 4. ATR-IR spectra of treated 5 wt% $CaCO_3$ filled EVA12 (T3).

Table 2. T-peel strength values (kN/m) of $CaCO_3$ filled EVA12/polychloroprene adhesive joints. Ageing conditions: 50°C/95 % relative humidity/72 hours.

Condition	**EVA-T1**	**EVA-T2**	**EVA-T3**	**EVA-T4**
Untreated	0.3	0.7	0.7	0.7
Corona discharge-treated	1.4	2.0	1.7	2.5
Untreated+aged	0.4	0.4	0.4	0.5
Corona discharge-treated + aged	2.4	3.2	2.1	3.1

In order to more precisely assess the loci of failure of the joints, the failed surfaces obtained after peel tests were analyzed using ATR-IR spectroscopy and SEM. The ATR-IR spectra of the failed surfaces show an adhesion failure in the untreated $CaCO_3$ filled EVA12 (Figure 5). The ATR-IR spectrum of one of the surfaces corresponds to the adhesive (A surface) and the other to the filled EVA12 (P surface). After corona discharge treatment the locus of failure of the joints changes. A mixed failure (adhesion + cohesion in EVA) is obtained, because some bands of filled EVA (C-O and C=O bands at 1238 and 1739 cm^{-1}; CH_2 and CH_3 bands at 2847, 2915 and 1464 cm^{-1}) can be distinguished in the ATR-IR spectrum of the A surface. This locus of failure is similar in all joints but becomes more cohesive in the T4 filled EVA material (containing the $CaCO_3$ with higher particle size).

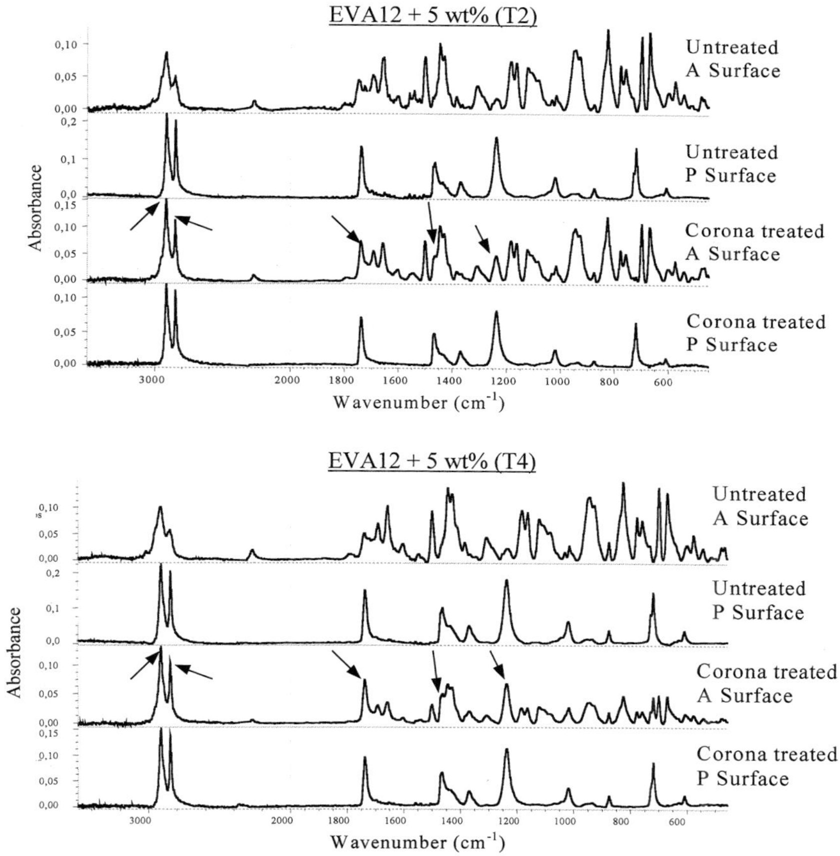

Figure 5. ATR-IR spectra of the failed surfaces obtained after peel tests of the joints produced with T2 and T4 filled EVA12.

The SEM micrographs of the failed surfaces of the $CaCO_3$ filled EVA/polychloroprene adhesive joints (Figure 6.a) show more ripped surfaces when treated with corona discharge, irrelevant to the particle size of the filler. All the SEM micrographs of the A surfaces show $CaCO_3$ particles which are transferred from the P surface during peel test (Figure 6.b), the transfer is more important in the EVA12 filled with higher particle size calcium carbonate, T4. The SEM micrographs of the P surfaces of the corona discharge treated samples show some EVA filaments after the peel test, as a consequence of the higher strength required to peel out the joint.

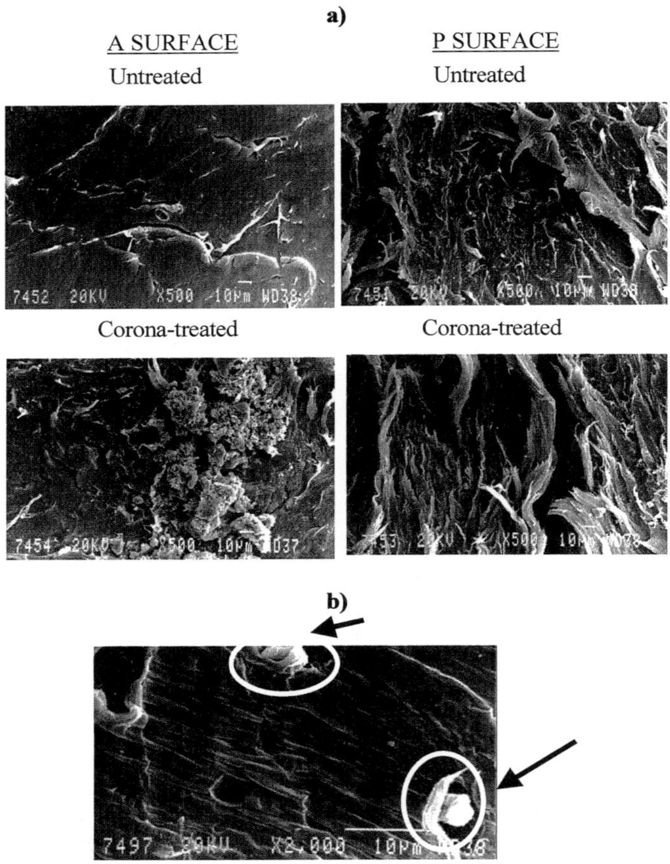

Figure 6. a) SEM micrographs of the failed surfaces obtained after peel test of the joint produced with T4 $CaCO_3$ filled EVA12; b) SEM micrograph of a zoomed part of the failed A surface.

Durability of the adhesive joints was studied by accelerated ageing tests. Table 2 shows that the ageing at 50°C and 95 % relative humidity for 72 hours produces an increase in peel strength values in the adhesive joints produced with the corona discharge-treated filled EVA12. The peel strength values are higher for joints produced with T2 and T4 $CaCO_3$ filled EVA12 treated with corona discharge.

An adhesion failure of the aged untreated filled EVA12/polychloroprene adhesive joints was obtained, as shown by the ATR-IR spectra of the failed surfaces (Figure 7), but a cohesive failure is obtained for all the aged corona discharge-treated joints, independently of the particle size of the filler. This may likely indicates that the reaction of the polychloroprene with the polyisocyanate is completed during ageing, allowing an increase in the cohesive strength of the adhesive and thus a greater peel strength value is obtained.

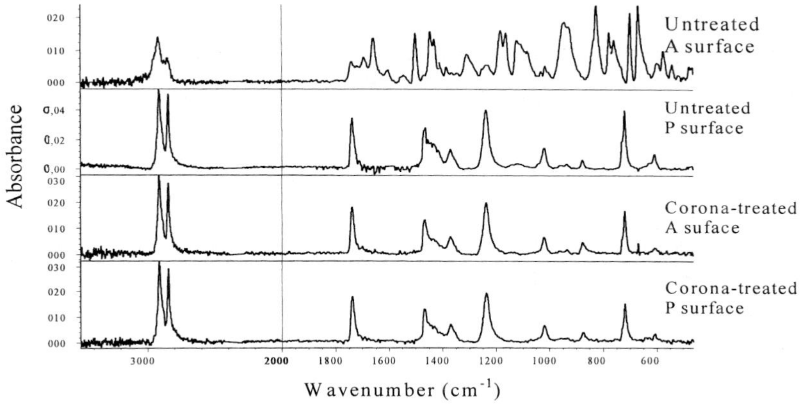

Figure 7. ATR-IR spectra of the failed surfaces obtained after peel tests of the aged adhesive joint of T4 $CaCO_3$ filled EVA12.

Conclusions

The corona discharge treatment exposed $CaCO_3$ particles to the surface of EVA12, and it seemed that the dispersion of the filler was better with T4 carbonate (higher particle size). The corona discharge treatment increased the peel strength values and higher values were obtained for the EVA containing $CaCO_3$ with higher particle size (4 μm). In general, a mixed (adhesion + cohesion in EVA) failure in the corona discharge-treated materials was produced, but the failure was more cohesive in the joint of T4 filled EVA12. The ageing of the joints increases the adhesion of the corona discharge treated materials.

Acknowledgements

Authors thank S.A. Reverté and Composan Adhesivos S.A. for supplying the calcium carbonate and the polychloroprene adhesive, respectively. Financial support by the Spanish Research Agency MICYT (project PTR1995-0578-CT) is also acknowledged.

[1] *Cargas minerales*, Ed: J.M. Martín Martínez and A.C Orgilés Barceló, University of Alicante, **1997,** Chap 1.
[2] Polymeric Materials Encyclopedia, Ed. J.C. Salomone, CRC Press, New York vol 4, **1996,** p.2365.
[3] M. Strobel, C. Dunatov, J.M. Strobel, C.S. Lyons, S.J. Perron and M.C. Morgen, *J. Adhesion Sci. Technol.* **1989,** *3,* 321.
[4] K.L. Mittal (Ed), *Polymer Surface Modification: Relevance to adhesion*, VSP, Utrecht **1996**.
[5] I. Sutherland, R.P. Popat and D.M. Brewis, *Journal of Adhesion,* **1994,** *46,* 79-88.
[6] A. Martínez-García. *PhD doctoral Thesis,* University of Alicante, **1999,** 228-249.

Macromol. Symp. **2005**, *221*, 33-41

Effect of the Amount of Calcium Carbonate as Filler on the Rheological and Adhesion Properties of a Water-Based Polyurethane Dispersion

Ana Belén Muñoz-Milán, Mª Ángeles Pérez-Limiñana, Francisca Arán-Aís, Ana Torró-Palau, A. César Orgilés-Barceló*

INESCOP, Footwear Research Institute. 03600 Elda (Alicante), Spain
E-mail: adhesivos@inescop.es

Summary: In this study, the properties of water-based adhesives based on a polyurethane ionomer (PUD) and a micronised $CaCO_3$ as filler was analysed. Different amounts of a micronised $CaCO_3$ (5-25 wt%) were added to water-based polyurethane (PUD) adhesive formulations in order to reduce its relatively high cost. The addition of a micronised calcium carbonate filler increased the viscosity, the storage and loss moduli of the PUD adhesives, and imparted pseudoplasticity and thixotropy, more noticeably for the adhesive with the highest calcium carbonate content. The creation of acid-base interactions seemed to be responsible for the improvement in the rheological properties of the PUD adhesives containing $CaCO_3$ as filler. On the other hand, the addition of $CaCO_3$ filler might deteriorate the adhesion properties to PUD adhesives.

Keywords: adhesive; calcium carbonate; dispersion; filler; rheology; water-based polyurethane

Introduction

Water-based polyurethane adhesives are an interesting alternative to the current solvent-based polyurethane adhesives used in the footwear industry because of their lack of flammability and toxicity that complies with the restrictive environmental legislation (Directive 99/13/EC) with respect to the emission of organic solvents into the atmosphere. Water-based polyurethane dispersions (PUD) offer an efficient alternative to solvent-based adhesives, the one most commonly used in the upper-to-sole bond in the footwear industry [1]. However, their relatively high cost is one of their drawbacks. Generally, micronised calcium carbonates are mainly used to reduce formulation costs, while to a certain extent controlling the viscosity during the preparation and application of several adhesives [2-3]. For this reason, the effect of incorporating different amounts of a micronised calcium carbonate on the properties of a water-based polyurethane dispersion was studied with the aim of improving some properties and reducing its cost.

 DOI: 10.1002/masy.200550304

Experimental

Materials. In order to perform a systematic study of the influence of adding $CaCO_3$ on the properties of water-based polyurethane adhesives, PUD adhesives with different percentage of a micronised $CaCO_3$ were prepared. The PUD adhesives contain the polyurethane dispersion and amounts between 0 and 25 wt% micronised $CaCO_3$ with respect to the total adhesive mass. An aliphatic, aqueous thermoplastic polyurethane dispersion based on highly crystalline polycaprolactone polyurethane (Quilastic DEP-170) provided by Merquinsa, Barcelona (Spain) was used to prepare the filled PUD adhesives. It is an anionic ionomer polyurethane dispersion with a mean particle size between 0,1-0,3 μm. The micronised $CaCO_3$ was provided by S.A. Reverté, Tarragona (Spain), and some of its characteristics are included in Table 1 [4].

Table 1. Some characteristics of the micronised calcium carbonate.

PROPERTY	**VALUE**
Size (μm) (95%)	<0,6
BET specific surface area (m^2/g)	14,5
Oil absorption (g/100g)	20
Humidity (%)	<0,1
Apparent density (g/mL)	0,50
Compacted apparent density (g/mL)	0,80
pH	8,8

The filled PUD adhesives were prepared by the following procedure: the micronised $CaCO_3$ was added to the PUD dispersion gradually and the mixture was stirred in a Cowles mixer at 850 rpm for 45 min, after which a homogeneous dispersion was obtained. Some measurements were carried out using water free PUD films prepared by placing about 50 mL of adhesive dispersions in a mould and allowing the water to evaporate slowly under ambient conditions.

Experimental techniques

Determination of solids content. The solids content of the PUD adhesives were determined according to the European Standard EN-827.

pH measurements. pH values were measured using a pH-meter with a silver reference electrode model HI 8418 (Hanna Instruments), at 20°C and according to the European Standard EN ISO 4045.

Rheology of the adhesive solutions. The viscosity, pseudoplasticity and thixotropy as a function of the shear stress of filled PUD adhesives were analysed in a Bohlin CS-50 rheometer. The measurements were carried out at 25°C in the rotational mode using concentric cylinders type C25. 25 mL of PUD adhesive dispersions were used in the measurements.

Rheology of water-free PUD films. The viscoelastic properties of the water-free PUD films were determined in a Bohlin CS-50 rheometer by using a plate-plate geometry. Oscillatory experiments were performed by melting the adhesive film at 70°C which is the reactivation temperature of the PUD adhesives used. The target strain was 0,005, and the frequency was varied between 0,01 and 30 Hz. All the experiments results were obtained in the region of linear viscoelasticity.

T-peel strength tests. Adhesion properties were assessed through T-peel strength on split leather/PUD adhesive/SBR rubber joints. SBR rubber was roughed and subsequently halogenated (2 wt%TCI/MEK), while the split leather was gently roughed and an adhesive primer was applied to it (European Standard EN 1392). After carrying out the surface treatment, the PUD adhesive was applied to each strip surface and left to dry for 30 min. The PUD adhesive film was heated to 70°C using infrared radiation in order to facilitate the interlocking of the polyurethane chains on the split leather and SBR surfaces during the joint formation. The strips were then put into contact and a pressure of 0,8 MPa was immediately applied for 10 s to achieve a suitable joint. T-peel strength was measured using an Instron model 1011 (peel rate = 0,1 m/min). The values obtained were the average of five test replicates (standard deviation was less than 5%). The green (immediate) peel strength was measured 30 s after the joint formation. The evolution in T-peel strength was followed at different times (until 72 h) after joint formation. In order to analyse the durability of the adhesive joints, they were aged at 70°C and 95% relative humidity for 72 h before the T-peel test was carried out.

Results and Discussion

Table 2 shows some properties (pH and solids content) of the aqueous polyurethane dispersions containing different amounts of $CaCO_3$ filler. An increase in the pH value of the dispersion occurs as the $CaCO_3$ content increases, as a consequence of the basic nature of the $CaCO_3$ (pH=8,8).

Waterborne polyurethane adhesives consist in polymer ionomer particles stabilizing in water due to ionic groups located on the particles surface (that act as an internal emulsifier chemically attached to the polyurethane structure) which generate an electrical double layer. Electrostatic forces between particles can be modified by varying the pH because of the double layer size of the particles is modified due to an increase of the electrostatic forces between particles, in agreement with the DLVO (Deryaguin-Landau-Verwey-Overbeek) theory [5-8].

Table 2. pH values and solids content of the water-based polyurethane adhesives containing different amounts of a micronised $CaCO_3$ filler.

$CaCO_3$ content (wt%)	**0**	**5**	**10**	**15**	**20**	**25**
pH	7,9	8,3	8,4	8,4	8,5	8,5
Solids content (wt%)	50,7	53,7	56,0	58,60	61,4	63,6

Unlike solvent-based adhesives, the viscosity of a water-based polyurethane adhesives is not dependent on the molar mass of the polymer but mainly on the solids content and mean particle size of the dispersion and by the presence of other components in the formulation. Therefore, the variation of the viscosity of the dispersions as a function of $CaCO_3$ content was performed. Figure 1 shows the flow curves of the adhesives with different content in $CaCO_3$. The addition of $CaCO_3$ filler produces an increase in viscosity due to the increase in solids content (Table 2), more noticeable for the adhesives containing 15, 20 and 25 wt% $CaCO_3$. Moreover, all the adhesives show a pseudoplastic behaviour, i.e. there is a decrease in viscosity as the shear stress increases, more markedly as the filler content increases [9]. The pseudoplastic behaviour of the filled PUD dispersions is due to a rupture of the network between the filler and the polyurethane particles and/or water, which becomes gradually broken by increasing the shear stress, giving a decrease in viscosity. In this case, this network could be ascribed to the creation of culombic forces or acid-base interactions [10].

When a dispersion is subjected to a shear stress, the polymer particles are orientated into layers, which can flow easily, causing a decrease in viscosity. The increase in viscosity and the pseudoplastic behaviour produced by breaking of the internal network can be ascribed to the formation of aggregates between particles which is more favoured as the $CaCO_3$ content increases. On the other hand, the presence of larger particles (filler) located between the smallest ones (polymer particles) may inhibit the orientation of particles restricting their ability to flow and produces an increase in viscosity.

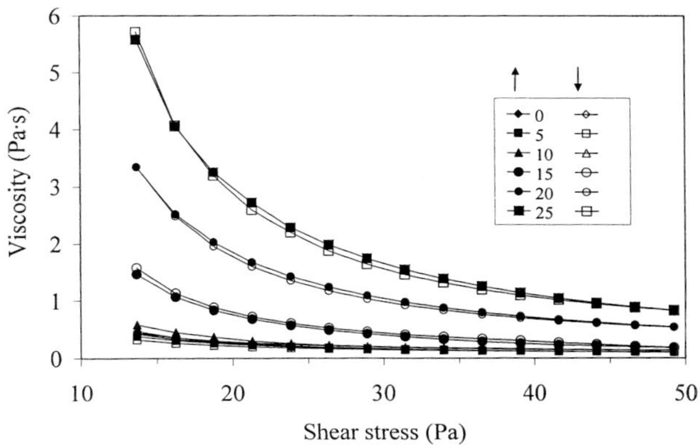

Figure 1. Variation of the viscosity for the filled polyurethane dispersions as a function of the shear stress. (↑ ↓ Increase and decrease in shear rate).

The yield point of the filled solutions (i.e. the minimum shear at which the polyurethane adhesive starts to flow) can be obtained by applying the Casson model to the rheological data:

$$\sigma^{1/2} = \sigma_o^{\ 1/2} + (\eta_\infty \cdot D)^{1/2}$$

where σ is the shear stress, D is the shear rate, σ_o is the yield point, and η_∞ is the viscosity at infinite rate. The yield point values of the filled PUD adhesives are included in Table 3. The increase in $CaCO_3$ content produces an increase in yield point stress from 15 wt% on. An increase in yield point stress involves a greater interaction between filler, water and polyurethane particles.

Table 3. Yield stress (σ_o) of filled PUD adhesives (obtained from Casson model).

$CaCO_3$ content (wt%)	**0**	**5**	**10**	**15**	**20**	**25**
σ_o (Pa)	7,58	6,62	6,99	10,32	9,21	9,21

Figure 2 shows the evolution of the storage modulus (G') as a function of the frequency of the filled water-free polyurethane films, obtained at 70ºC. The addition of $CaCO_3$ filler to water-based polyurethane adhesives produces an increase in the modulus up to a 20 wt%, more noticeably in the low frequency region of the curves, indicating the resistance of the filler-PUD blend to flow at these frequencies. The improvement in the rheological properties implies the strengthening of the structure up to a $CaCO_3$ content of 20 wt% which can be ascribed to the creation of culombic forces or acid-base interactions. The mechanical properties of a composite depend on the degree of filler dispersion in the polyurethane matrix. In an acid or base polymer matrix, the filler dispersion is favoured due to acid-base interactions [11]. V. Kovacevic *et al.* [12] found that the polar nature of $CaCO_3$ affects the adhesion between the PVAc and the filler. A certain degree of similar interactions can be expected in PUD/$CaCO_3$ composites. The bonding between $CaCO_3$ and PUD phases in a dispersed system can be explained by the establishment of coordinate bonds between the anionic groups of the ionomer polyurethane and the Ca^{2+} ions [12].

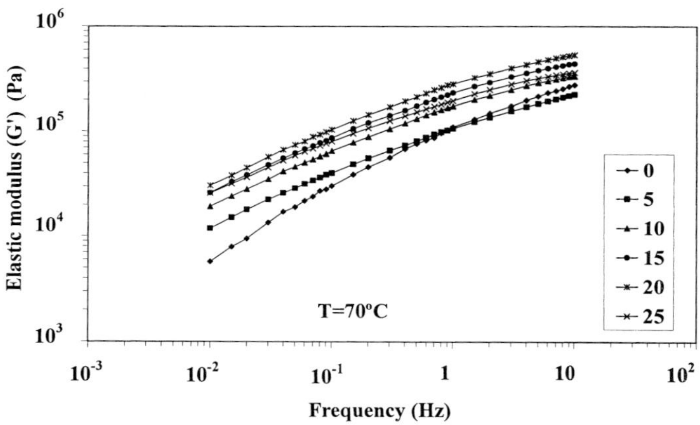

Figure 2. Variation of the storage modulus (G') of the filled PUD films as a function of the frequency.

For filler content higher than 20 wt%, a decrease in the storage modulus is produced. It could be ascribed to the fact that a high content of $CaCO_3$ as filler inhibits the interactions among polyurethane chains, which are the main responsible of the polyurethane structure and therefore of their rheological properties [13].

On the other hand, the PUD films (Figure 3) show, at a given frequency, a change from viscous to elastic properties due to the entanglement of polyurethane chains. It means that there is a cross-over between the storage (G') and loss (G'') moduli. The addition of micronised $CaCO_3$ filler to polyurethane produces a displacement of the cross-over to lower frequencies, regardless of the filler content (Table 4). Therefore, the addition of a micronised $CaCO_3$ to polyurethane improves their rheological properties as it strengthens the structure by avoiding the flow of the polymer chains.

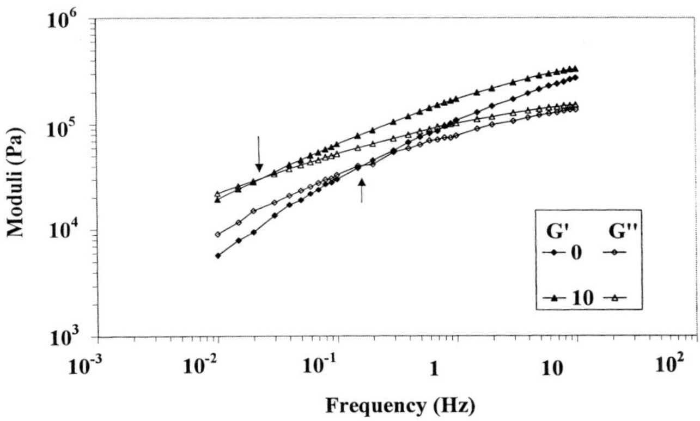

Figure 3. Variation of the elastic (G') and viscous (G'') moduli of the polyurethane films with 0 and 10 wt% of micronised calcium carbonate as a function of the frequency.

Table 4. Frequency ($\omega_{cross\text{-}over}$) and modulus ($G_{cross\text{-}over}$) values in the cross-over for the different adhesives.

$CaCO_3$ content (wt%)	**0**	**5**	**10**	**15**	**20**	**25**
$G_{cross\text{-}over}$ (Pa)	$5{,}5\cdot10^4$	$1{,}8\cdot10^4$	$2{,}8\cdot10^4$	$3{,}8\cdot10^4$	$4{,}5\cdot10^4$	$3{,}2\cdot10^4$
$\omega_{cross\text{-}over}$ (Hz)	0,30	0,02	0,02	0,02	0,02	0,02

Finally, the influence of the $CaCO_3$ content on adhesion properties of PUD adhesives was assessed by means of T-peel tests. Figure 4 shows the evolution of the T-peel strength values in leather/PUD adhesive/SBR joints as a function of the $CaCO_3$ content.

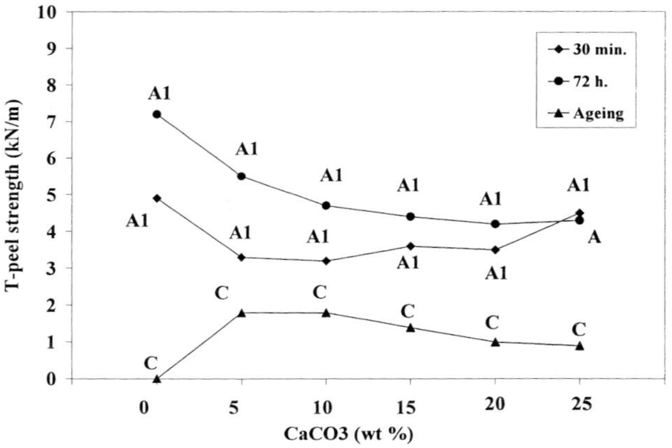

Figure 4. T-peel strength values at different times after formation of leather/polyurethane adhesive/SBR joints as a function of the micronised $CaCO_3$ content. Locus of failure: C= Cohesive failure in the adhesive; A1: Adhesion failure between the PUD adhesive and the leather.

The initial adhesion (30 min) for all joints produced with the adhesives containing filler is, in general, slightly lower than the one produced with the adhesive without filler. Moreover, 72 hours after joint formation the peel strength values decrease as the amount of filler increases, but the peel strength values are always higher than the initial adhesion as a consequence of the crystallisation of the polyurethane chains. An adhesion failure to the leather (A1) was always obtained. It can be ascribed to the higher viscosity of the PUD adhesives containing $CaCO_3$ as filler which may prevent the penetration into the pores of the leather, facilitating the adhesion failure. Moreover, the improvement in the reological properties causes more rigid and fragile films in comparison to the adhesive without filler. On the other hand, ageing produces a decrease in peel strength values in all joints, the decrease is less important if the adhesive contains $CaCO_3$. According to the literature, the calcium carbonate as filler forms a shield that prevent ageing process [14]. During ageing the hydrolysis of polyurethane chains is produced causing a cohesive failure in the adhesive (C).

Conclusions

The addition of a micronised calcium carbonate as filler increased the viscosity, the storage and loss moduli of the PUD adhesives and imparted pseudoplasticity and thixotropy, more noticeably for the adhesive with the highest calcium carbonate content. The creation of acid-base interactions seemed to be responsible for the improvement in the rheological properties of the PUD adhesives containing $CaCO_3$ as filler. However, the addition of $CaCO_3$ produces a decrease in peel strength for leather/polyurethane adhesive/SBR joints more noticeable as the filler content increases.

Acknowledgements

Partial financial support from MCYT (project No. MAT 2002-03117) is acknowledged. Authors thank Reverté S.A. (Spain) for providing the $CaCO_3$ and Merquinsa (Spain) for providing the water-based polyurethane dispersion used in this study.

[1] J.C. Fernández García, M. Cobo Ulloa, A. Torró Palau, A.C. Orgilés Barceló. *Lederpiel* **1998,** 72 January-February.
[2] J. Sepulcre-Guilabert, T. P. Ferrándiz-Gómez, J.M. Martín-Martínez. *J. Adh. Sci. Technol.*, **2001**, 15, 187.
[3] J. Sepulcre-Guilabert, T.P. Ferrándiz-Gómez , J.M. Martín-Martínez. *Macromolecular Symposia* **2001**, 169, 185.
[4] "*Technical information*". Reverté S.A, Bellvei, Tarragona 1999.
[5] "*Encyclopedia of emulsion technology. Basic theory. Vol. 1*" P. Becher, Ed., Marcel Dekker, Inc., New York 1983.
[6] L. Chan Young, K. Jin Woong, S. Kyung Do. *J. Mater. Sci.,***1999**, 34, 5343.
[7]. "*Handbook of adhesives*". 3rd ed., I. Skeit, Ed., Van Nostrand Reinhold, New York 1999.
[8] "*Chemistry of Pigments and Fillers*", D. H. Solomon, D.G. Hawthorne, Eds., Krieger Publishing , New York, Chichester 1976.
[9] "*Viscoelastic Properties of Polymers*", Ferry, Ed., Wiley, New York, Chichester 1976.
[10] D. Dietrich in: "*Advances in urethane ionomers*". H.X. Xiao, K.C. Frish, Eds., Technomic, Lancaster, USA 1995. Chapter 1.
[11] M. Ciganez, M. Dressler. *Polym. Composites,* **1989,** 10, 194.
[12] V. Kovacevic, S. Lucic, D. Hace, Z. Cerovecki. *J. Adh. Sci. Techol.*, **1996**, 10, 12, 1273.
[13] "*Szycher's handbook of polyurethanes*". M. Szycher, Ed, CRS Press. Boca Ratón, Florida 1999.
[14] A. Baker, L. Koller, E. Kummer in "*Handbook of filler for plastics*". 2nd ed., H.S. Katzs, J.V. Milewski, Eds. Van Nostrand Reinhold, New Jersey 1987.

Improved Adhesion of UV Treated EVA Copolymer by Adding Small Amounts of Fillers

María Dolores Landete-Ruiz, José Miguel Martín-Martínez*

Adhesion and Adhesives Laboratory, Department of Inorganic Chemistry, University of Alicante, 03080 Alicante, Spain
E-mail: md.landete@ua.es

Summary: The influence of adding small amounts of calcium carbonate or sepiolite fillers on the adhesion properties of UV surface-treated ethylene vinyl acetate copolymer (EVA) was studied. In this paper, it has been demonstrated that the addition of 3-7 wt% filler enhanced the extent of the surface modifications (wettability, surface chemistry, roughness) and increased the adhesion of EVA. These effects were more marked when 7 wt% sepiolite filler was added to EVA, likely due to its higher aspect ratio.

Keywords: EVA; fillers; infrared spectroscopy; peel strength; UV treatment

Introduction

Ethylene vinyl acetate (EVA) copolymers are commonly used as sole materials in the footwear and toy industries due to their chemical resistance, light weight, flexibility and toughness. However, adhesion of EVA is poor due to its low surface energy, and, therefore, a surface treatment is necessary. Several treatments (corona discharge, low-pressure oxygen or air plasmas) resulted in the formation of polar moieties and increased surface energy, which led to enhanced adhesion of different polymers [1], including polyolefins. On the other hand, the UV-ozone treatment has been used to increase the wettability of poly(ethylene terephtalate) (PET), polyethylene (PE), and polypropylene (PP) [2,3]. Furthermore, it has been shown [4] that the UV radiation treatment improved the wettability and adhesion of unfilled EVA-polyethylene foams and injection moulded EVA [5].

On the other hand, polyolefins are usually loaded with fillers (calcium carbonate, silica, silicates) to increase their mechanical properties and/or to reduce costs. The way in which the fillers affect the effectiveness of the surface modifications and adhesion of polyolefins has not been already considered in the existing literature. In this study, the surface modifications and adhesion of an EVA filled with small amounts of calcium carbonate or sepiolite were considered.

 DOI: 10.1002/masy.200550305

Calcium carbonate is the most widely used filler for plastics, paper, paints and coatings (milled grades), and it is also employed in emulsion paints, printing inks, cigarette paper, pharmaceutical preparations, and toothpaste (precipitated grades). As examples, the addition of calcium carbonate improves the impact strength and inhibits the migration of additives in PVC [6]. In PE and PP films, calcium carbonate filler acts as anticaking agent [7].

Sepiolite ($Si_{12}O_{30}Mg_8(H_2O)_4(OH)_4 \cdot 8H_2O$) is a hydrated magnesium aluminum silicate, with layered talc-like structure composed of ribbons with two sheets of tetrahedral silica units linked by means of oxygen atoms to a central octahedral sheet of magnesium cations. Sepiolite is less used as filler because its deposits are relatively rare in comparison with the other industrial clays, and Spain is the largest producer (it accounts for about 95% of the world's annual production) [8]. Sepiolite is used in several applications owing to its fibrous structure, relatively high aspect ratio and high porosity [9]. The major uses of sepiolite are in drilling muds, paints, liquid detergents, adhesives, car polishing, filler in plastics and rubber, blend compatibilizer, cosmetics and also as a binder for pelletized animal feed. PP filled with sepiolite shows improved bending and tensile strength [10]. Sepiolite was also used to improve the adhesion of solvent-based polyurethane adhesives [11,12].

Previous studies [13] showed that, after plasma treatment, similar degree of surface modification and adhesion improvement was obtained in filled EVA with 3 to 20 wt% $CaCO_3$. Optimum performance was obtained in EVA containing 3 wt% $CaCO_3$. On the other hand, it has been shown that the adhesion of EVA can be improved by addition of 3 to10 wt% of sepiolite [14], showing a slight increase in adhesion by increasing the amount of sepiolite. However, the addition of 10 wt% sepiolite imparted stiffness to EVA, so a loading of 7 wt% was considered as optimum.

Therefore, in this study small amounts of calcium carbonate (3 wt%) and sepiolite (7 wt%) were used as inert and reinforcement fillers for one EVA copolymer containing 12 wt% vinyl acetate – EVA12. The surface modifications produced by UV-ozone treatment and the adhesion properties of unfilled and filled EVA12 were compared.

Experimental

The EVA12 pellets were supplied by Repsol Química, S.A. (Santander, Spain). This material has a melt flow index of 0.6 g/10 min and a Shore D hardness of 40 [15].

EVA12 was mixed with 3 wt% $CaCO_3$ – EVA12C - or 7 wt% sepiolite – EVA12S - in a extruder. The unfilled and filled EVAs were moulded using a Margarit JSW injection machine to obtain test samples of 150 x 60 x 2 mm^3. These samples were cut into 150 x 30 x 2 mm^3 for adhesion tests.

Sepiolite was supplied by Tolsa, S.A. (Madrid, Spain). This filler contains needle shaped particles (2 μm length, 50-100 Å thick). The natural micronized calcium carbonate was supplied by S.A. REVERTÉ (Tarragona, Spain) and has a mean particle size of 1.9 μm. Some properties of the fillers are given in Table 1.

Table 1. Some properties of sepiolite and calcium carbonate [16] .

Property	$CaCO_3$	Sepiolite
Density (g/l)	400 (uncompacted) – 700 (compacted)	60
Moisture content (wt%)	<0.1	8
Melting point (°C)	---	1550
pH	8.8	8.7
DOP (dioctyl phthalate) absorption (g DOP/100g)*	27	250
BET surface area (N_2/77K) (m^2/g)	14.5	270

*ISO 787-5 standard.

Adhesion of UV-treated EVAs were obtained from joints made with a two-component solvent-based polyurethane adhesive based on Desmocoll 540 pellets supplied by Bayer AG (Leverkusen, Germany). This polyurethane has a high crystallization rate and a short open time. Adhesive solutions were prepared by dissolving 18 wt% solid Desmocoll 540 pellets in butanone in a laboratory mixer using a Cowles stirrer (3 h, 850 rpm, 25°C). 5 wt% Desmodur RFE polyisocyanate (Bayer AG, Leverkusen, Germany) was added immediately before adhesive application.

The UV radiation source was a low-pressure vapour grid mercury lamp, manufactured by American Ultraviolet (USA). The lamp works at the wavelength of 254 nm, which provided a radiation intensity of 10 mW/cm^2 taken at a distance of 2.54 cm from the lamp. Samples were treated on a platform mounted 2.0 cm under the UV lamp. The length of the treatment was varied between 1 and 5 minutes.

The changes in the surface properties of EVAs due to UV radiation treatment were measured using different experimental techniques.

Contact angle measurements. They were carried out in a thermostated chamber (25 °C) of a Ramé Hart 100 goniometer. The chamber was previously saturated with the vapour of the test liquid at 25 °C (bidistilled and deionized water) for at least 10 minutes before placing a 4 μl drop of the test liquid on the EVA surfaces. Advancing contact angles of deionized and doubly distilled water were measured on just UV-treated EVAs. The experimental error was ±2 degrees.

ATR-IR spectroscopy. The use of attenuated total reflectance technique of IR (ATR-IR) examination allows a very thin surface layer (2 to 5 μm depth, depending on prism nature and experimental conditions) to be examine by clamping a specimen to a prism from which the IR radiation is reflected. The infra-red spectra of the treated samples were obtained in a Nicolet FTIR 5DXB spectrometer provided with a thallium bromo-iodide crystal; the incident angle was 45° and 200 scans were collected and averaged at a resolution of 4 cm^{-1}. Under the experimental conditions used in this study, the ATR-IR spectroscopy allows to analyze about 2 μm of the treated EVA surface.

X-ray photoelectron spectroscopy (XPS). A VG Scientific Microtech Multilab spectrometer with a MgK_α X-ray source (1253.6 eV) operating at 15 keV and 300 W was used; a 45° incidence angle was used. The analysis was performed on 5 x 2 mm pieces of EVA at a residual pressure below 10^{-8} Torr. For each sample, a survey scan encompassing the region 0-1200 eV was first obtained. Multiplex scanning of all observed photopeaks in the survey scan were carried out in a 20 eV range. Binding energies of all photopeaks were referenced to the C 1s photopeak position for C-C and C-H (hydrocarbons) species at 285.0 eV. 1.8 eV was used as FWDH for the carbon peak.

Scanning electron microscopy (SEM). The morphological modifications produced on the treated EVA surfaces were analyzed using a *JEOL JSM-840 SEM* microscope. The samples were gold-coated and the energy of the electron beam was 20 kV.

T-peel strength. Adhesive joints using two similarly treated EVAs were produced. Immediately after UV treatment, about 0.8 ml of polyurethane adhesive solution was applied to each EVA test specimen using a brush. After allowing the solvent to evaporate for 20 min, the dried polyurethane adhesive films on the surface-treated EVA were melted for 10 seconds at 100ºC using infrared radiation. These specimens were then placed in contact and a pressure of 0.8 MPa was applied for 10 s to achieve a suitable joint. The T-peel strength was measured using an Instron 4411 instrument (peel rate = 100 mm/min). The values obtained were the average of three experimental replicate determinations (standard deviation was less than 10%). The adhesive joints were kept at 25ºC and 50% relative humidity before undergoing the T-peel test. Adhesion was measured 72 hours after joint formation.

Results and Discussion

The water advancing contact angle values on EVA surfaces (Figure 1) decrease by increasing the length of the UV treatment. Unfilled and filled EVA12 show similar trend but somewhat lower values were obtained in the UV treated filled EVAs. Therefore, the UV treatment increases the wettability of EVA12, more markedly when a small amount of filler is added, irrelevant of its nature.

The decrease in the contact angle values can be ascribed to modifications in the surface chemistry and/or morphology of EVAs.

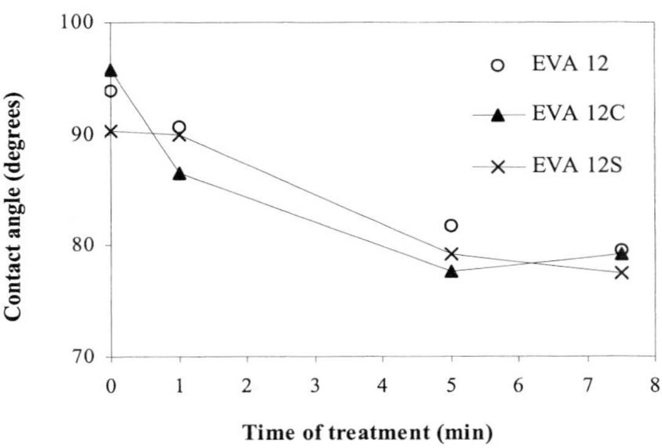

Figure 1. Water (25ºC) advancing contact angles on EVAs treated with UV radiation.

Figure 2 shows the ATR-IR spectra of the as-received and UV treated EVAs. The ATR-IR spectra of the as-received EVAs show the typical bands of vinyl acetate at 1739, 1242, 1023 and 606 cm^{-1}, and ethylene at 2919, 2846, 1467, 1375, and 725 cm^{-1}. The filled EVAs also show the typical bands of the fillers : Bands at 880 and 1437 cm^{-1} due to the CO_3^{2-} group (EVA12C), and bands at 1017 and 480 cm^{-1} due to Si-O group (EVA 12S). The UV treatment decreases the relative intensity and increases the width of the C=O band at 1739 cm^{-1}, indicating the formation of new oxygen-containing moieties on the treated EVA surfaces. However, because of about 2 μm thick EVA is analyzed by ATR-IR spectroscopy, a more sensitive surface analysis technique was used to determine the surface modifications produced on the EVA surfaces by UV treatment.

(a)

(b)

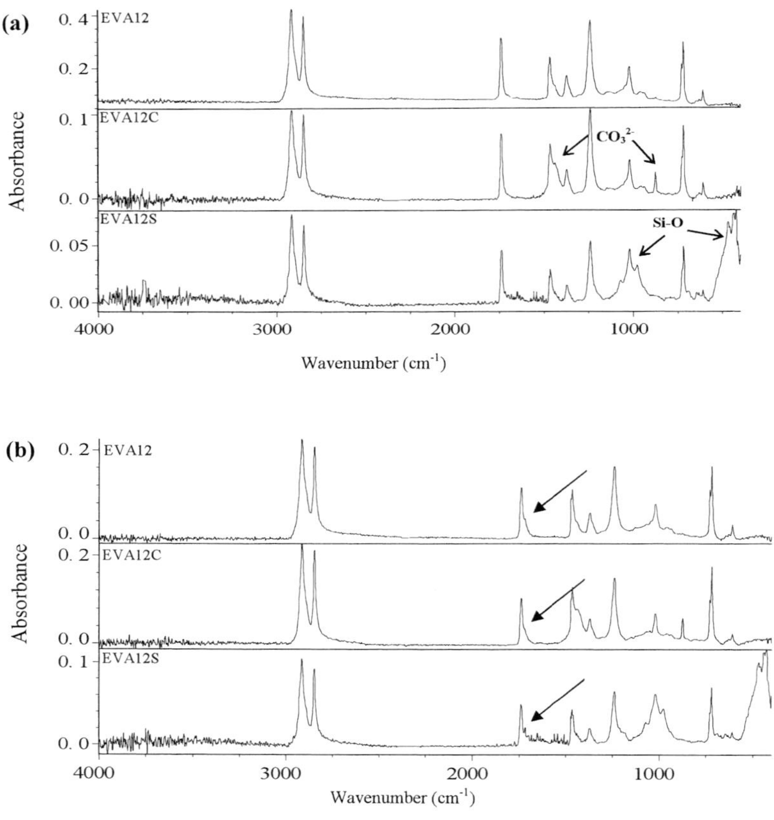

Figure 2. ATR-IR spectra of the (a) as-received and (b) UV-treated (5 min) EVAs.

Table 2 shows the elemental composition on the as-received and UV-treated EVAs obtained using XPS. As-received EVA12 is mainly composed of carbon and a small amount of oxygen. EVA12S contains small amounts of Si and Mg due to the presence of sepiolite particles on the surface, but Ca is not present on the surface of EVA12C. Because XPS analyses about 5 nm surface and Ca is shown in the ATR-IR spectra of the as-received EVA12C (about 2 μm depth is analyzed), the $CaCO_3$ particles on the surface must be covered by a thin layer of polymer.

Table 2. Elemental composition on the as-received and UV-treated (5 min) EVA surfaces. XPS experiments.

Sample	Element	As-received	UV treated	
			1 min	5 min
EVA 12	C	95.2	95.2	92.8
	O	4.8	4.8	7.2
	O/C	**0.05**	**0.05**	**0.08**
EVA 12C	C	95.2	95.0	92.1
	O	4.8	5.0	7.9
	Ca	---	---	---
	O/C	**0.05**	**0.05**	**0.09**
EVA 12S	C	91.0	85.8	81.7
	O	7.1	10.4	14.2
	Si	1.8	3.7	3.9
	Mg	0.1	0.1	0.2
	O/C	**0.08**	**0.12**	**0.17**

The UV treatment for 5 min produces an increase in oxygen and consequently in the O/C ratio on the treated EVA surfaces. For a shorter UV treatment an increase in oxygen content is only found on the EVA12S surface. Furthermore, the extent of the surface modifications are more marked in the filled EVA12, mainly when sepiolite is added, indicating an effect of the filler content and nature in the modifications produced on the EVA surface by treatment with UV radiation.

Figure 3 shows the percentages of carbon species obtained from C1s curve fitting of as-received and 5 min UV-treated EVAs. As-received EVAs show the typical bands corresponding to EVA: 285.0 eV (C-C, C-H), 285.6 eV (C^*H_3-(C=O)-O), 286.6 eV (HC^*-O-(C=O)-), and 289.2 eV ((C^*=O)-O). The UV treatment decreases the amount of C-C and C-H species (285.0 eV), an slight increase in C-O moieties is produced, and new C=O moieties (288.0 eV) are created. The presence of calcium carbonate or sepiolite produces a more noticeable decrease in C-H moieties on UV-treated EVA12C and EVA12S, and a more marked increase in C-O groups in EVA12S.

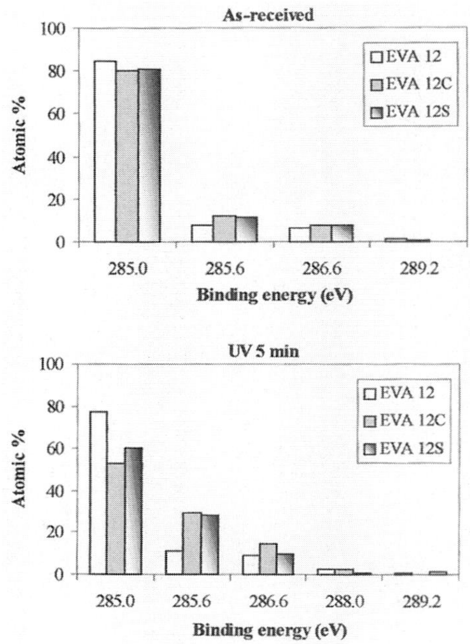

Figure 3. Distribution of C1s species on (a) as-received and (b) UV treated (5 min) EVA surfaces. XPS experiments.

The UV treatment also modifies the morphology of EVAs. SEM micrographs (Figure 4) show some cracks on the surface of treated EVA12. As-received filled EVAs (EVA12C and EVA12S) show a more heterogeneous surface (small particles of $CaCO_3$ and sepiolite appear on the surface) than the unfilled EVA12, and the UV treatment generates more degraded surfaces in a non-uniform way.

As a consequence of the chemical and morphological modifications on the EVAs produced by UV-treatment, improved adhesion in EVA 12/polyurethane + 5 wt% polyisocyanate adhesive joints is produced (Figure 5).

The joints produced with the as-received EVA12 show very low peel strength values due to the poor wettability and the non-polar nature of the copolymer, irrelevant to the presence or not of fillers. However, the UV treatment of EVAs increases the peel strength values, which is in agreement with the improvement in wettability, and the creation of roughness and surface chemistry. The UV treated sepiolite filled EVA12 show the highest adhesion improvement, in agreement with the greater extent of the surface modifications produced by the treatment.

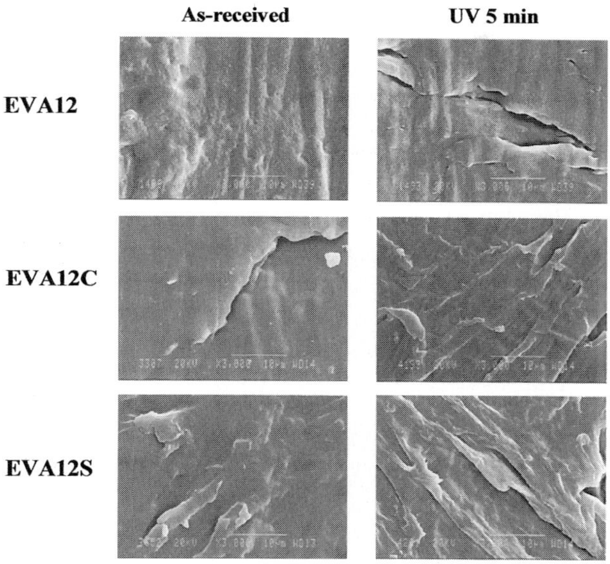

Figure 4. SEM micrographs of as-received and UV treated (5 min) EVAs.

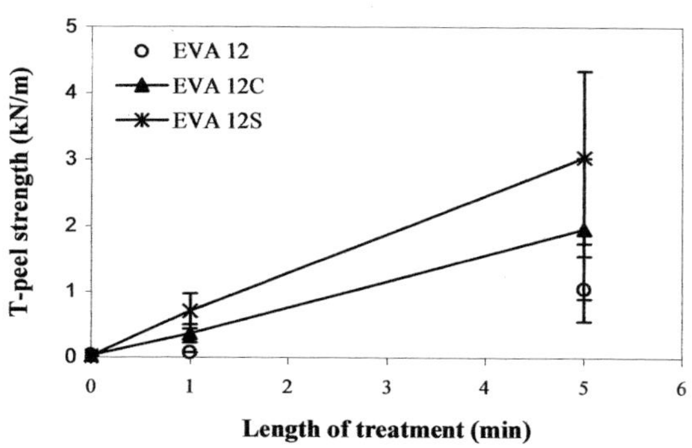

Figure 5. T-peel strength of as-received and UV treated EVA 12/polyurethane + 5 wt% polyisocyanate adhesive joints.

Conclusions

The UV treatment increases the adhesion of EVAs to polyurethane adhesive due to improved wettability, and the creation of roughness and C-O and C=O moieties. The presence of small amounts of $CaCO_3$ or sepiolite fillers in EVA 12 enhanced the extent of the surface modifications, and better performance was obtained when 7 wt% sepiolite is used as filler for EVA12.

Acknowledgments

Financial support from the Spanish Research Council (MCYT project PTR1995-0578-CT) is acknowledged. M.D. Landete-Ruiz thanks the Generalitat Valenciana for granted her PhD Thesis. The authors also thank Repsol Química S.A., Tolsa, S.A, S.A. REVERTÉ, AIJU and Bayer AG for supplying some of the materials used in this study.

[1] *"Polymer Surface Modification: Relevance to Adhesion, Volume 2"*, K.L. Mittal, Ed., VSP, Utrecht 2000, pp. 3-27, 377-392.
[2] M.J. Walzak, S. Flynn, R. Foerch, J.M. Hill, E. Karbashewski, A. Lin, M. Strobel, *J. Adhesion Sci. Technol.* **1995**, *9*, 1229.
[3] I. Mathieson, R.H. Bradley, *Int. J. Adhesion and Adhesives* **1996**, *16*, 29.
[4] M.D. Landete-Ruiz MD, J.A. Martínez-Díez, M.A. Rodríguez-Pérez, J.A. de Saja, J.M Martín-Martínez, *J. Adhesion Sci. Technol.* **2002**, *16*, 1073.
[5] M.D. Landete-Ruiz, J.M. Martín-Martínez, *Int. J. Adhesion and Adhesives*, in press.
[6] X.L. Xie, Q.X. Liu, R.K.Y. Li, X.P. Zhou, Q.X. Zhang, Z.Z. Yu, Y.W. Mai, *Polymer* **2004**, *45*, 6665.
[7] W.C.J. Zuiderduin, C. Westzaan, J. Huétink, R.J. Gaymans, *Polymer* **2003**, *44*, 261.
[8] H. Murray, *Industrial Clays Case Study,* International Institute for Environment and Development and World Business Council for Sustainable Development **2002**, *64*, 1.
[9] J.L. Martín-Vivaldi, R.H.S. Robertson, in: *"Optical Investigation of Clays"*, J.A. Gard, Ed., Mineralogical Society, London 1971, p. 225.
[10] A. Linares, E. Morales, M.C. Ojeda, J.L. Acosta, *Angew. Makromol. Chem.* **1986**, *147*, 41.
[11] A. Torró-Palau, J.C. Fernández-García, A.C. Orgilés-Barceló, M.M. Pastor-Blas, J.M. Martín-Martínez, *Int. J. Adhesion and Adhesives* **1997**, *17*, 111.
[12] A. Torró-Palau, PhD thesis, University of Alicante, Alicante, Spain 1997.
[13] C.M. Cepeda-Jiménez, PhD thesis, University of Alicante, Alicante, Spain 2001.
[14] A. Martínez-García, PhD thesis, University of Alicante, Alicante, Spain 1999.
[15] Technical information of Repsol, *Copolymer EVA Alcudia,* Repsol Química, Santander, Spain 1996.
[16] Technical Information of S.A. REVERTÉ, Bellvei, Tarragona, Spain 1997; Technical information of TOLSA S.A, Madrid, Spain 1996.

Polypropylene-Clay Nanocomposites: Comparison of Different Layered Silicates

Andreas Leuteritz, Doris Pospiech, Bernd Kretzschmar, Meike Willeke, Dieter Jehnichen, Ulrike Jentzsch, Karina Grundke, Andreas Janke*

Institute of Polymer Research Dresden, P. O. Box 12 04 11, 01005 Dresden, Germany
E-mail: leuteritz@ipfdd.de

Summary: In former studies of the preparation of polypropylene(PP)-clay nanocomposites, different types of layered silicates were used. However, the obtained results were not comparable due to different preparation conditions and types of silicates. The aim of this work was the investigation of the influence of different layered silicates on the properties of the resulting nanocomposites. FT-IR-spectra, SAXS, TEM micrographs, elemental analysis, mechanical properties and surface tension measurements were used for the comparison of the four different layered silicates under investigation.

Keywords: clay; layered silicates; nanocomposites; polypropylene; surface tension measurements

Introduction

In the last years, polymer nanocomposites containing layered silicates have gained great attention due to their academic and industrial importance[1–6]. They have shown a significant improvement in mechanical, thermal, especially flame-retardant, and barrier properties with small amounts of inorganic fillers compared to conventional filler composites. Improved properties were obtained with loadings of only 1–10 wt% of modified layered silicates. Polypropylene as one of the most widely used and low cost polyolefin is very interesting as matrix polymer for such nanocomposites. Before preparing the nanocomposite, the clay has to be transferred to a more organophilic form to obtain a better interfacial adhesion to the nonpolar polypropylene. Clay modification is generally achieved by ion exchange reactions of the intergallery sodium cations of the layered silicates by organophilic cations (like alkyl ammonium ions). To produce polypropylene layered silicate nanocomposites by a melt mixing method, maleic anhydride grafted polypropylene (PP-g-MA) has to be used as compatibilizer [7,8]. The aim of the present paper is to evaluate the influence of four different layered silicates on the polypropylene

 DOI: 10.1002/masy.200550306

nanocomposite properties. Therefore, a fixed standard composition of the nanocomposite as well as a standard preparation technology was used in order to obtain an expressive result.

Experimental

Four different types of layered silicates were used to produce organophilic modified clays: Nanofil 757 (Germany; CEC = 75 mmol/100 g), denoted MM(1), and Nanofil 918 (Turkey; CEC = 100–110 mmol/100 g), denoted MM(2), both types are products of Süd-Chemie AG (Germany). The second one is a natural clay with Ca-cations in the intergallery, which have to be exchanged by Na-cations in order to produce a silicate fully exchangeable by organic cations like alkylammonium compounds (activation) in aqueous media. Additionally, Cloisite Na^+ (CEC = 95 mmol/100 g) from Southern Clay (USA), denoted MM(3), and Somasif ME100 (CEC = 70 – 80 mmol/100 g) from CO-OP Chemicals (Japan), denoted MM(4), were used. These four layered silicates were modified with dimethyldistearylammonium chloride (DSQ) from Clariant (Genamin DSAC) according to a standardized method described elsewhere [9]. The corresponding nanocomposites [NC(1)–NC(4)] were produced in a batch technology via a master-batch of organoclay and PP-g-MA (Polybond 3200 from Crompton). The batch concentration is calculated on the basis of the ash content of the organoclay intending to produce a formulation containing 7.5 wt% PP-g-MA and 5 wt% inorganic residue[10] in the nanocomposite after dilution with PP-homopolymer (Moplen HF500N from Basell). The density of pure montmorillonite and hectorite in the sodium form is about 2,0 – 2,7 g cm^{-3} depending on the content of water surrounding the sodium ions and differs from cation exchanged form. The density of the organoclays varies due to different amounts of DSQ with respect to the CEC. In order to make the nanocomposites comparable, we decided to focus on the inorganic residue, giving a more precise similarity in the nanocomposite on volume and mass fraction of the filler used, taking into account the more or less "molecular" level of dimensions, rather to use a fixed weight fraction of the organoclay, which would definitely lead to different volume fractions in the nanocomposite.

Instrumental

FT-IR-Spectra were taken on a Bruker IFS66V. Clay materials were measured either as powder, in reflection mode, or as KBr disks. SAXS data were collected on a KRATKY compact camera

(AntonPaar Graz, Austria), and TEM-pictures were made using ultrathin cuts with a Philips CM 20. Elemental analysis was performed with an Elementar VARIO EL CHNS analyzer. Batches of the modified silicates with PP-g-MA were prepared on a cokneader MKS 30-18 (Buss, Switzerland) followed by compounding with PP (200-220 °C; 300 rpm; 10 kg/h) on a co-rotating twin screw extruder ZE 25 (Berstorff, Germany) to obtain the nanocomposite with the focused clay content. Mechanical analysis was done according to CAMPUS conditions. The inorganic residue was determined gravimetrically after burning in a microwave oven at 550 °C/10 min.
The modified organoclay samples were characterized by FT-IR-spectroscopy, SAXS, elemental analysis and a method to determine the surface tension reported earlier [11,12]. Tensile bars were made by injection moulding from the corresponding nanocomposites and were characterized by mechanical analysis, SAXS and TEM.

Results and Discussion

To prepare polypropylene based nanocomposites four different layered silicates MM(1)–MM(4) were used. Figure 1 shows for better visibility a cutout of the IR-spectra of MM(1)–MM(4). There are only insignificant differences in the IR-spectra due to the different amounts of not exchangeable ions M^{II+}/M^{III+} like Fe and Mg cations. These cations are responsible for the negative charge of the platelets, which is compensated by the exchangeable sodium ions in the intergallery. The synthetic hectorite MM(4) used for comparison is a three-layer silicate produced by heating talc and Na_2SiF_6 for several hours in an electric furnace at high temperatures (CO-OP Chemicals, Japan)[13]. Therefore, in the IR-spectra of MM(4) the Si–OH stretching band in the region of 3600 cm^{-1} is missing, which is the difference in the part of the IR spectra not shown. Beside those bands, different amounts of quartz (SiO_2) can be found in the IR spectra of the four different clay materials MM(1)–MM(4) (see Figure 1).

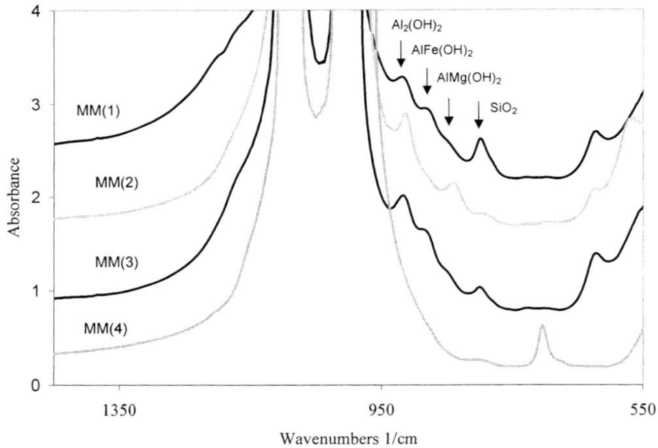

Fig. 1: FT-IR-Spectra of MM(1)–MM(4) silicates in the region of 550–1400 cm^{-1}

MM(1)–MM(4) were modified with DSQ in a stoichiometric ratio of 1/1 taking into account their different cation exchange capacity (CEC). The exact amount of DSQ in the modified clays DSQ MM(1)–DSQ MM(4) was obtained by elemental analysis, see Table 1. The given cation exchange ratio n/CEC was determined from the carbon value of the elemental analysis and is close to the theoretical value of 1.00 (100%).

Table 1: Elemental analysis of cation exchanged layered silicates and surface tension γ_{sv}

Sample	Carbon content [%]	Nitrogen content [%]	n / CEC	Surface tension γ_{sv} [mJ/m^2]
Unmodified MM	-	-	-	44.2
DSQ MM(1)	24.88	0.673	1.02	27.5
DSQ MM(2)	29.97	0.872	1.01	27.5
DSQ MM(3)	29.47	0.834	1.04	26.5
DSQ MM(4)	24.48	0.630	0.99	26.5

The surface tension γ_{sv} of the modified clays DSQ MM(1)–DSQ MM(4) measured according to method given in reference [11] are also given in Table 1. There are only neglected differences in the surface tension values of DSQ MM(1)–DSQ MM(4) - 26.5–27.5 mJ/m^2. But the aim of the modification to obtain a more hydrophobic silicate powder can be realized as reflected in the great difference in the surface tension of modified with respect to unmodified layered silicate. The surface tension of polypropylene determined in melt at 185 °C by using axisymmetric droplet shape analysis was about 20 mJ/m^2 [14]. Consequently, there is still a gap existing between the surface tension of modified clay and polymer matrix, which makes a compatibilizer necessary.
The *d*-spacings of the unmodified layered silicates are about 0.9 – 1.2 nm and result from sodium ions surrounded by different amounts of water molecules. After modification with DSQ the observed interlayer distances increase compared to the pristine silicate *d*-spacings. As shown in Figure 2, the four different layered silicates DSQ MM(1)–DSQ MM(4) differ in the interlayer distances reaching from about 2.7–3.2 nm. As expected, the clay MM(1) with the lowest cation exchange capacity shows the smallest *d*-spacing in the modified form.

The interlayer distance obtained from DSQ MM(4) is hardly comparable with those of the three clay samples DSQ MM(1)–DSQ MM(3), because of the differences in the layered silicates: MM(1)–MM(3) are natural montmorillonite, belonging to the family of dioctahedral smectites, and MM(4) is a synthetic hectorite belonging to the family of trioctahedral smectites. The interlayer distances from the batch composites DSQ MM(1) PB–DSQ MM(4) PB remain nearly the same compared to the original modified silicates, concluding, that there is no intercalation of polymer into the galleries. The prepared composites NC(1)–NC(4) are best described as partly exfoliated. The TEM images of NC(1)–NC(4) reflect, that the layered silicates in the nanocomposites are exfoliated to some extent, see Figure 3. Both, stacks of multilayer and isolated exfoliated platelets can be observed. The low-magnification TEM image (see Figure 3 left) shows that the modified layered silicate NC(2) is well dispersed throughout the polymer matrix.

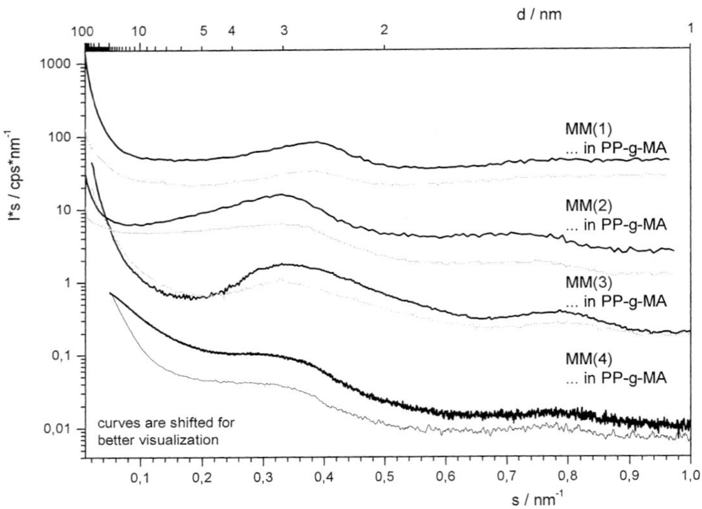

Fig. 2: SAXS curves for DSQ modified MM(1)–MM(4) and for the clay/PP-g-MA batches [DSQ MM(1) PB–DSQ MM(4) PB] to determine interlayer spacing distances (d)

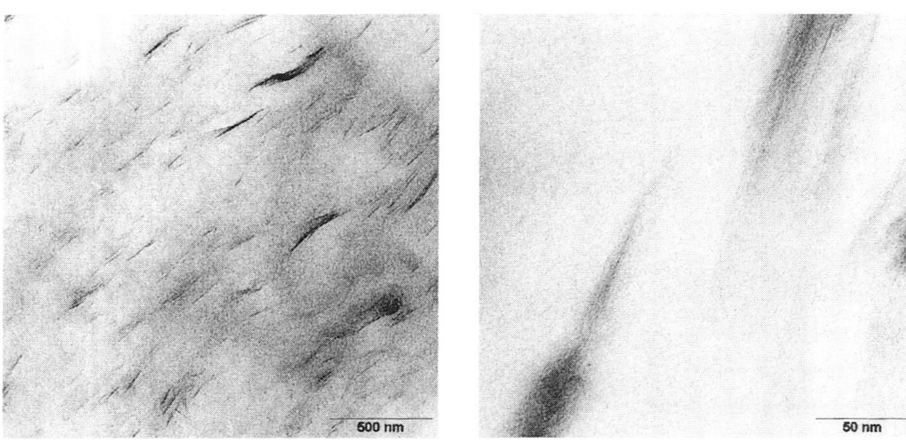

Fig. 3: TEM-micrographs of DSQ modified MM(2) in PP/PP-g-MA (left: low magnification; right: high magnification)

There exists a complex relationship between the properties of the layered silicates and the properties of the corresponding nanocomposites. Only the nanocomposites obtained from the natural clays MM(1)–MM(3) are comparable to each other due to the differences in the composition of the layered silicates.

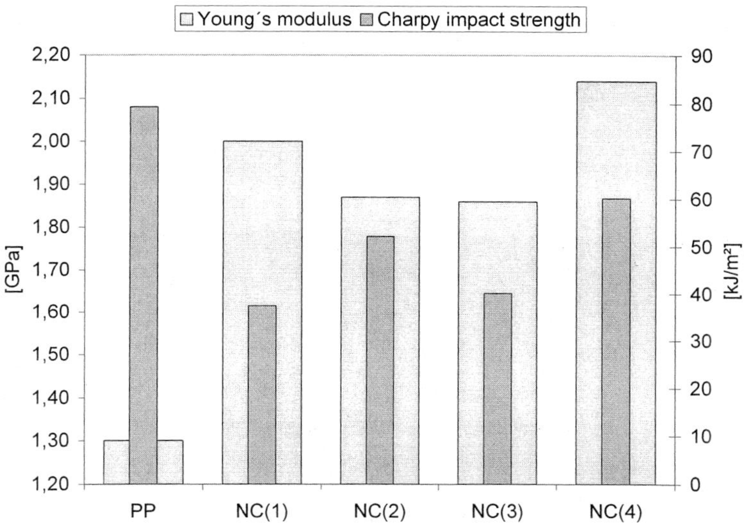

Fig. 4: Young's modulus and impact strength of NC(1)–NC(4) in comparison to PP

The experiments concerning the mechanical properties (see Figure 4 and Table 2) show the influence of the cation exchange capacity and the different amounts of organic molecules on the impact strength and Young´s modulus. The highest impact strength was found in the nanocomposite NC(2) with the clay with the highest CEC, the lowest in NC(1) with the clay having the lowest CEC. The differences in the notched impact strength and the tensile strength are inferior, but show the same general trend. Unfortunately, there is at present no possibility to accurately quantify the degree of exfoliation and therefore it is not clear, if the differences in mechanical properties result from different aspect ratios of the distributed clays, due to agglomerates and/or to different primary particle sizes. According to Morgan and Gilman[15] it is necessary to analyze SAXS-curves and TEM micrographs to get an idea of the clay-platelet dispersion. On the other hand, new rheological experiments leading to the shear thinning factor give only a qualitative degree of exfoliation within the layered silicate[16]. But for a correct analysis the mean size and size distribution of different layered silicates and their agglomerates would be necessary.

Table 2: Mechanical properties of nanocomposites based on DSQ modified layered silicates

Sample	Residue [%]	σ_Y [MPa]	ε_Y [%]	E_t [GPa]	a_{cU+23} [kJ/m^2]	a_{cA+23} [kJ/m^2]	HDT B [°C]
PP	-	30.5	9.5	1.33	79.2	3.9	85
NC(1)	5.1	35.1	6.8	2.00	37.4	3.6	111
NC(2)	5.2	35.2	7.4	1.87	52.1	4.2	105
NC(3)	4.7	34.3	7.8	1.86	40.1	3.6	104
NC(4)	4.8	36.2	7.0	2.14	60.1	4.3	109

Conclusion

In this work nanocomposites of four different layered silicates in polypropylene were compared. Three natural clays (montmorillonite) with different cation exchange capacities (CEC) and a synthetic hectorite were used. The nanocomposites were prepared in a batch technology with a focus of 7.5 wt.-% PP-g-MA and 5 wt.-% inorganic residue. The mechanical properties show the influence of the CEC on the strength and the modulus, when comparable clays (natural clays MM(1)-MM(3)) are used. MM(1) with the lowest CEC shows also the lowest impact strength in the corresponding nanocomposite whereas the highest impact strength is found in the composite NC(2) with the clay MM(2) having the highest CEC. The obtained results from NC(3) were intermediate. Although different amounts of alkylammonium ions were found in the modified clay samples due to their different CEC, the surface tension measurements show nearly the same results. Therefore, the surface tension is surprisingly more influenced by the differences in the organic modifiers[17] than by the cation exchange capacity of the layered silicates. The influence of the sizes of the platelets on the strength and modulus is at present hardly to quantify because of the difficulties in a suitable model and analytical methods. For a better understanding of the characteristics of the layered silicates in the polymer matrix further experiments have to be done.

Acknowledgement

The authors would like to thank Dr. T. Engelhardt (Süd-Chemie AG), L. Schäfer (Clariant GmbH) and the Federal Ministry of Education and Research (BMBF-Grant 03N8626) for supporting this research.

[1] P. C. LeBaron, Z. Wang, T. J. Pinnavaia, *Appl. Clay Sci.* **1999**, *15*, 11–29.
[2] M. Alexandre, P. Dubois, *Mat. Sci. Eng.* **2000**, 12, 1–63.
[3] R. Mülhaupt, T. Engelhardt, N. Schall, *Kunststoffe* **2001**, *91* (*10*), 178–190.
[4] E. Manias, A. Touny, L. Wu, K. Strawhecker, B. Lu, T. C. Chung, *Chem. Mater.* **2001**, *13*, 3516–3523.
[5] D. Schmidt, D. Shah, E. P. Giannelis, *Current Opinion in Solid State and Mater. Sci.* **2002**, *6*, 205–212.
[6] A. Leuteritz, D. Pospiech, B. Kretzschmar, M. Willeke, D. Jehnichen, U. Jentzsch, K. Grundke, A. Janke, *Adv. Eng. Mater.* **2003**, *5* (*9*), 678–681.
[7] P. Reichert, H. Nitz, S. Klinke, R. Brandsch, R. Thomann, R. Mülhaupt, Macromol. Mater. Eng. **2000**, 275, 8-17.
[8] N. Hasegawa, H. Okamoto, M. Kato, A. Usuki, *J. Applied Polym. Sci.*, **2000,** *78*, 1918.
[9] US-Pat.-No. 2,966,509; **1960**.
[10] This ratio was chosen due to an unpublished study.
[11] K. Grundke, A. Augsburg, *J. Adhesion Sci. Technol.* **2000**, *14* (*5*), 765–775.
[12] K. Grundke, T. Bogumil, T. Gietzelt, H.-J. Jacobasch, D. Y. Kwok, A. W. Neumann, *Progr. Colloid Polym. Sci.* **1996**, *101*, 58–68.
[13] *Synthetic Mica, Micromica*, CO-OP Chemicals Co., Ltd., Tokyo, Japan.
[14] P. Nygard, K. Grundke, E. Mäder, *J. Adhesion Sci. Technol.* **2002**, *16* (*13*), 1781–1808.
[15] A. B. Morgan, J. W. Gilman, J. Applied Polymer Science **2003**, 87, 1329-1338.
[16] Wagener, R., Reisinger, T. J. G., Polymer **2003**, 44, 7513-7518.
[17] A. Leuteritz, D. Pospiech, B. Kretzschmar, D. Jehnichen, U. Jentzsch, A. Janke, K. Grundke, Materials Week 2002, Proceedings - Frankfurt: Werkstoffinformationsges., 2003 - ISBN 3-88355-314.

Macromol. Symp. **2005**, *221*, 63-73

Poly(propylene)/PET/Undecyl Ammonium Montmorillonite Nanocomposites. Synthesis and Characterization

José Ignacio Velasco,[1] Mónica Ardanuy,[1] Lourdes Miralles,[2] Silvia Ortiz,[1] Maria Lluisa Maspoch,[1] Miguel Sánchez-Soto,[1] Orlando Santana*[1]

[1] Centre Català del Plàstic (CCP), Universitat Politècnica de Catalunya (UPC), Edifici Vapor Universitari, Carrer Colom 114, E-08222 Terrassa (Barcelona), Spain
Fax: (34) 937841827; E-mail: jose.ignacio.velasco@upc.es
[2] Laboratorio de Investigación en Formaciones Salinas, Departamento de Geoquímica, Petrología y Prospección Geológica, Universidad de Barcelona, Zona Universitaria de Pedralbes, E-08028 Barcelona, Spain

Summary: A natural bentonite rich in calcium montmorillonite (CaMMT) was initially purified and ion-exchanged to obtain sodium montmorillonite (NaMMT). Both clays were organophillised by cationic exchange reaction with undecyl-ammonium chloride, and characterized. Isotactic poly(propylene) (PP) was melt-compounded with both the unmodified and the organophilic montmorillonites. The hybrids produced have been characterized structurally, thermally and mechanically. Maleic anhydride-grafted PP (MAH-g-PP) was used as compatibilizer in some of the formulations. Homologous series of hybrids were also synthesized employing blends of PP/PET and compared with those of the pure PP to investigate possible beneficial effects due to the presence of small amounts of PET on the microstructure and properties of this kind of materials. The analysis of the results indicates some extension of both macromolecules intercalation and clay particles exfoliation in the hybrids prepared with the organophilic montmorillonite. The hybrids prepared with compatibilized PET/PP blends were found to have a better nanostructure.

Keywords: blends; clay; nanocomposites; PET; PP

Introduction

The method called polymer melt intercalation was proposed by Giannelis to fabricate polymer/clay nanocomposites[1]. Nowadays, polypropylene/clay nanocomposites show great research interest[2-8] since this thermoplastic is one of the most widely used because of its good ratio between properties and cost, as well as due to its high versatility. Kawasumi et al.[9] used maleic anhydride modified PP oligomers as compatibilizer to prepare PP nanocomposite by melt blending. They found that the dispersibility of the clay layers depends on the miscibility of the PP maleic anhydride oligomers with the matrix PP matrix. Hasegawa et al.[4] reported a novel approach to prepare PP/octadecylamine modified montmorillonite using maleic anhydride-graft-

 DOI: 10.1002/masy.200550307

polypropylene as compatibilizer and they obtained that silicate layers were partially exfoliated and dispersed to monolayers. Gloaguen and Lefebvre[10] prepared by melt blending dispersion of organophilic clay in PP without compatibilizer and they found that polymer-clay interaction exists although no specific treatment was used to promote it. Nam et al.[11] prepared intercalated polypropylene/clay hybrid by using small amount of maleic anhydride groups as compatibilizer to achieve disordered structures. Recently, Liu and Wu[3] reported the synthesis of intercalated polypropylene/MMT nanocomposites by melt grafting.

In this work we have synthesized PP/clay hybrids using a natural bentonite, which was firstly purified to obtain sodium montmorillonite, and secondly treated with undecylammonium chloride. A blend of homopolymer PP with maleic anhydride-grafted PP was used as matrix to promote affinity with the organophilic montmorillonite. Moreover, taking in mind the affinity of some polar thermoplastics (i.e. polyesters, polyamides…) for high-polarity particle surfaces, homologous series of hybrids were synthesized using a blend of PP/PET as matrix, with the aim to investigate possible beneficial effects on the composite microstructure and properties due to the presence of PET.

Morphology and microstructure of the hybrids have been characterised by means of WAXS and TEM. Furthermore, thermal and mechanical properties as well as flame resistance of the nanocomposites have been evaluated.

Clay treatment

The bentonite (Minas de Gador, Spain) was treated physically in order to obtain a fine high-purity clay fraction (< 2 μm) before its chemical treatment. The procedure was performed in three steps: milling in a porcelain mortar, sieving, and centrifuging at 6000 rpm for 20 minutes. This fraction consists mainly of calcium montmorillonite.

Two types of cationic exchange reactions were performed on this clay. On one hand, sodium montmorillonite (NaMMT) was produced from CaMMT and, on the other hand, organophilic clays (CaOMMT and NaOMMT) were prepared by reaction of CaMMT and NaMMT respectively with undecyl ammonium chloride.

Each 30 g of CaMMT were first dispersed into 400 ml of 1 N solution of NaCl for two hours at room temperature. After centrifugation at 3000 rpm for 15 min the solid fraction was separated.

To assure complete ion exchange, that is, to obtain an homoionic NaMMT the same process was repeated five times. The product was washed several times with a mixture of deionised water and ethanol (1:1 by volume) until no chloride was detected by adding one drop of 0.1 N $AgNO_3$ solution. The purified clay was dried at 110°C and stored in a desiccator.

Organophilic clays were prepared using a similar method to that reported by Yano et al.[6]. In a 500 ml beaker 10 g of 11-aminoundecanoic acid, 5 ml of concentrated hydrochloric acid, and 100 ml of water were placed. The solution was heated at 80 °C. Moreover, 20 g of MMT was dispersed in 400 ml of water at 80 °C. The dispersion of MMT was added to the solution of undecyl ammonium chloride, and this mixture was stirred for 12 hours. A white precipitate was isolated by filtration, placed in a 500 ml beaker with 400 ml of hot water, and stirred for 1 hour. This process was repeated twice to remove the residue of ammonium salt. The final product was then filtered and dried at 110 °C, and stored in a desiccator.

Materials, compounding and specimens

A commercial grade (Repsol-IPF, Isplen PP050) of homopolymer PP was used as matrix to prepare the hybrids. A commercial compatibilizer (Eastman, Epolene G-3003) consisting of maleic anhydride-grafted PP copolymer (MA-g-PP) was added in some formulations. To investigate possible beneficial effects of the existence of poly(ethylene terephthalate) (PET) on the characteristics of these materials, analogous hybrids were synthesized using a blend of PP with PET (Catalana de Polímers, *Extrupet EW36*) and compared with those of pure PP.

Polypropylene/clay (PP/MMT) hybrids were prepared by a melt-compounding process performed in two stages. Firstly, highly filled PP/MMT and PP/OMMT compounds with nominal MMT concentration of 20 wt% were produced using a *Collin ZK-35* co-rotating twin-screw extruder (D = 25 mm; L/D = 36). Intensive dispersive mixing was assured by means of three kneading blocks inserted in the screw configuration. A barrel temperature profile was selected from 150 °C at the polymer feeding to 190°C at the die. Clay was fed in the extruder through a feeding port located at a distance of 12D from the polymer feeding. Vacuum devolatilization was applied at a distance of 24D and the screw speed was fixed at 300 rpm. Under these conditions, the melt temperature measured at the die was never higher than 200°C. A circular cross-section die of 3 mm diameter was employed. The extrudate was cooled in a water bath and pelletised. In a second extrusion process, performed under the same conditions, the clay concentration was reduced to nominal percentage of 4 wt% by dilution with the neat PP. Four polymer/clay hybrids were prepared: two hybrids containing pure PP and both kinds of clay (CaMMT and CaOMMT) called respectively PPMMT and PPOMMT, and their analogous containing 2 wt% of MA-g-PP as compatibilizer, called PMMMT and PMOMMT respectively. Table 1 shows their nominal composition by weight. The clay percentage was 4 wt% in all cases.

Square plaques (150 x 150 x 0.8 mm^3) and discs (diameter 80 mm, thickness 3 mm) were compression-moulded using a hot-plate press. The moulding cycle was optimised for each plaque geometry and material. The temperature was never higher than 190 °C. For burning test, the square plaques specimens were machined to 120 x 10 x 0.8 mm^3 dimensions. Standard tensile specimens (type IV, ASTM D-638) were also cut off. Square specimens (20 x 20 x 3 mm^3) were machined from the discs to determine the Vicat softening temperature and for WAXS and TEM analysis.

Testing procedure

Clay density was measured using an He pycnometer model AccuPyc 1330. The cationic exchange capacity (CEC) of the clays was determined using the following procedure: on one hand, 100 mg of clay were stirred for 10 minutes in 50 ml of ammonium acetate (pH 7.0) in order to exchange ions. It was left to stand 24 hours and, afterwards, the solution was analysed by atomic absorption spectroscopy (AAE) and atomic emission spectroscopy (AES) to quantify respectively the residual Ca^{2+} and Mg^{2+}, and K^+ and Na^+ ions.

Table 1. Composition of both unmodified and organophilic MMT-containing hybrids.

Unmodified MMT series:	PP-MMT	PM-MMT	BP-MMT	BMP-MMT
Modified MMT series:	PP-OMMT	PM-OMMT	BP-OMMT	BMP-OMMT
PP (wt.%)	96	94	91	89
MA-g-PP (wt. %)		2		2
PET (wt. %)			5	5

On the other hand, the clay was washed thoroughly with isopropyl alcohol to remove any excess of NH_4^+ ions. This was checked using a Nessler reactive. The sample was then treated with 30 ml of a NaCl 1N solution slightly acidified with 0.005N HCl to extract the NH_4^+ ions from the clay. This procedure was monitored by flux injection analysis (FIA).
Thermogravimetric analysis (TGA) was carried out on a Setaram TG-DTA92 thermoanalyser. The sample was heat from 20 °C to 900 °C at 1°C/min under oxygen atmosphere.
Wide angle X-ray scattering (WAXS) was performed using a Siemens D-500 diffractometer with CuK_α radiation (wavelength λ=0.154 nm), operating at 40 KV and 30 mV. The step size was 0.05° and the measuring time 5 s/step. The clay samples were prepared using a technique for aggregates orientation, and the polymer hybrids specimens were machined from the circular plaques. The clay interlayer basal spacing (BS) was evaluated through the (001) reflection of montmorillonite.
The melt flow index (MFI) of the PP/clay hybrids was measured at 230 °C and 2160 g. The Vicat softening temperature was determined at a heating rate of 120 °C/h with a load of 50 N.
Differential scanning calorimetry (DSC) was carried out in a Perkin Elmer Pyris-1 calorimeter to analyse the polymer thermal characteristics in the hybrids (typically 10 mg taken from the circular plaque). After erasing the thermal history (4 min at 220°C) the sample was cooled at 10 °C/min to 40 °C to observe the PP crystallization. The temperature was next raised from 40 to 220°C at 10 °C/min to observe the polymer melting signal. All tests were conducted under nitrogen atmosphere.
The silicate layers in the hybrids were observed by transmission electron microscopy (TEM) on ultrathin microtomed sections taken from the compression-moulded discs.
The mechanical properties of the hybrids were evaluated by tensile tests carried out at 1 mm/min, according to ASTM D-638 standard.
To analyse eventual flame retardancy effects, vertical burn test (UL-94) was carried out. The time from the ignition to the occurrence of the first burning drops (TFBD) was measured.

Results and Discussion

The CEC values of the clays have been compiled in Table 2. On one hand, it can be seen that the value obtained (95,1 meq/100 g) for the original montmorillonite (CaMMT) is concordant with values of usual MMT commercial grades employed in nanocomposites. On the other hand, the efficiency of the clay organophillisation process was high, as the low Na^+ and Ca^{2+} residues values (from AAS and AES) obtained for samples CaOMMT and NaOMMT reveal, which indicates that a high fraction of all metallic ions in the clay have been exchanged by organic ones. This result was confirmed by TGA. Besides the weight loss due to the adsorbed water and structural OH^-, a weight loss was observed in OMMT in the interval 180-300°C which should be related to the organic substance. This weight loss of 16 wt% was in accordance with a CEC value of 95 meq/100g.

Table 2. Some physicochemical characteristics of the clays.

	Density	Cationic Exchange Data		Basal Spacing
		AAS / AES	FIA	
	g/cm^3	meq/100 g	meq/100 g	Å
CaMMT	2.52	85.8	95.1	14.9
NaMMT	2.41	121.1	134.6	12.6
CaOMMT	2.18	11.2	93.3	16.6
NaOMMT	2.13	8.4	99.4	16.7

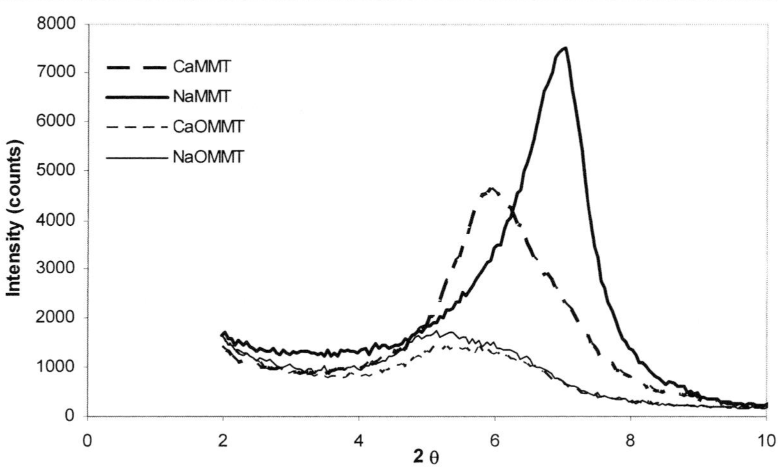

Figure 1. WAXS difractograms of the different clays.

The WAXS diffractograms of the original clay exhibited a peak at $2\theta = 5.93°$ corresponding to a basal spacing of 14.9 Å. Treatment with undecyl ammonium chloride produced a slight expansion to 16.6 Å (Fig. 1). The lower density values of the organophilic clays are explained due to this increase of the clay basal spacing. Owing to the fact that this value was similar for both organophilic clays, we decided to employ the original calcium-based clays (CaMMT and CaOMMT) to prepare the polymer/clay hybrids instead of NaMMT and NaOMMT.

A diffraction peak exists in the WAXS diffrcatograms of the hybrids. A better degree of delamination could be achieved in BMP-OMMT sample since its (001) intensity resulted very low (Fig. 2).

Poly(propylene) intercalation into the galleries of the organophilic clay seems to be evident after the TEM analysis (Fig. 3a). Furthermore, intercalation of PET molecules would be expected due to the presence of the undecanoic acid group into the organophilic clay galleries. It was observed that PET encapsulated clay platelets and particles due to its high polarity (Fig 3b).

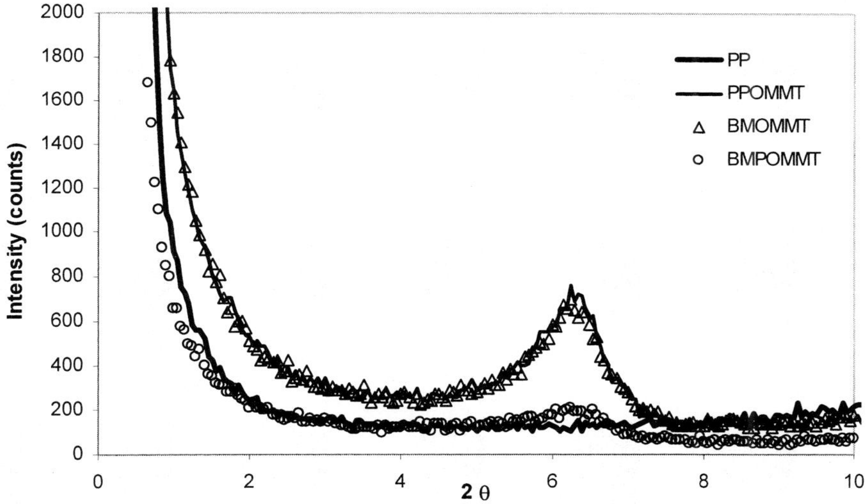

Figure 2. WAXS diffractograms of some selected hybrids.

The presence of organophilic clay promotes an increase in the PP Vicat softening temperature, which is especially relevant in the hybrid PPOMMT and in those containing PET (Table 3).

Moreover the hybrids PPOMMT and PMOMMT exhibits TFBD values notably higher than the rest of the samples, indicating a flame retardant effect of OMMT for poly(propylene).

Table 3. MFI, Vicat and TFBD values of all the samples.

	Melt Flow Index g/10 min	Vicat Softening Temperature °C	TFBD[a)] sec
PP	5.8	97.3	2
PP-MMT	3.9	101.3	3
PM-MMT	4.3	101.9	4
PP-OMMT	5.9	108.6	7.5
PM-OMMT	5.8	101.4	7
BP-MMT	3.9	97.1	2.5
BMP-MMT	4.4	108.6	3
BP-OMMT	4.1	103.4	3.5
BMP-OMMT	4.8	109.0	2.5

a) Time from ignition to the first burning drop (test geometry UL-94).

Table 4. Thermal and mechanical characteristics.

	T_m [a)] °C	T_c [a)] °C	X_m [b)] %	X_c [b)] %	E [c)] MPa	σ_{max} [d)] MPa	$\varepsilon_{\sigma max}$ [e)] %	ε_{break} [f)] %
PP	159.5	111.5	53.9	52.9	690	28.9	13.8	> 100
PP-MMT	161.4	117.1	56.5	55.8	820	31.1	12.5	> 100
PM-MMT	162.7	120.0	53.0	50.7	851	30.7	11.8	21
PP-OMMT	164.8	121.3	55.9	53.4	884	32.1	9.9	27
PM-OMMT	162.8	120.3	53.1	51.5	932	31.6	10.0	29
BP-MMT	165.1	123.1	53.8	52.4	876	26.8	6.4	6.4
BMP-MMT	164.0	123.7	54.8	52.3	869	30.2	8.0	11
BP-OMMT	165.3	126.6	49.1	49.2	879	29.8	8.5	80
BMP-OMMT	164.5	126.6	49.2	47.4	880	30.4	7.8	17

a) Melting and crystallization peaks temperatures. b) Crystallinity values from melting and crystallization signals. c) Young modulus. d) Tensile strength. e) Elongation at maximum stress. f) Elongation at break.

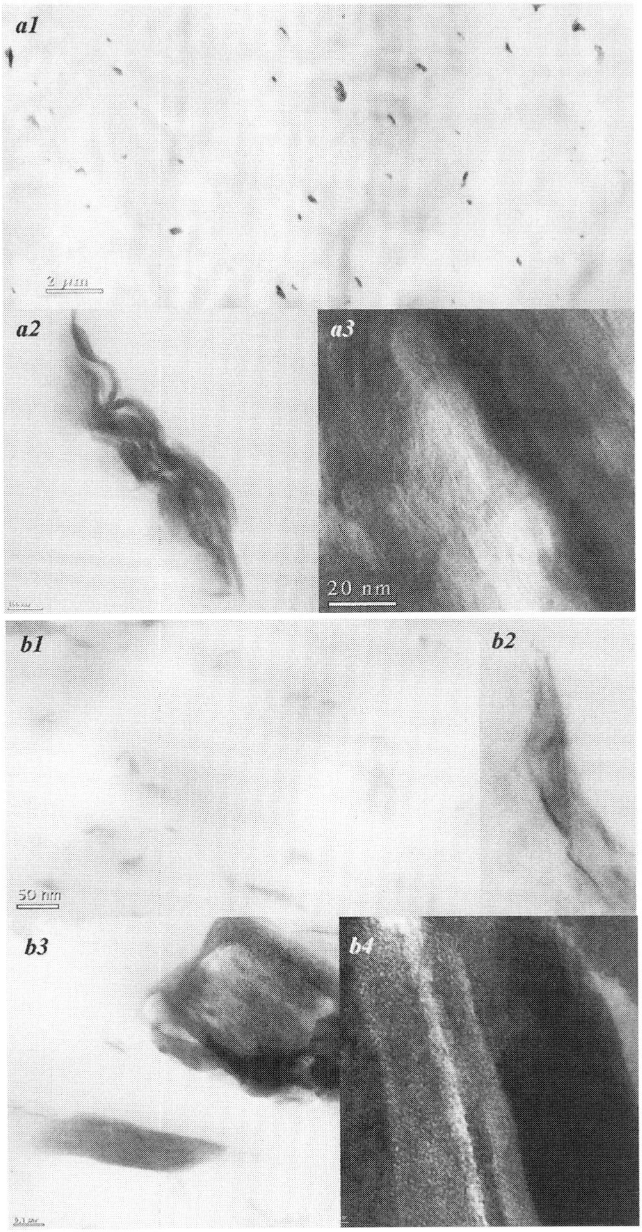

a1) Primary clay particles and agregates dispersed within PP matrix.

a2) Agregate of primary clay particles.

a3) Magnification of a primary clay particle.

b1) Exfoliated clay platelets and PET domains dispersed within PP matrix.

b2) Individual clay platelets surrounded by PET.

b3) Primary clay particles surrounded by PET.

b4) Magnification of a primary clay particle.

Figure 3. TEM analysis of a) PP/OMMT and b) PP/PET/OMMT hybrids.

However, it is remarkable that the TFBD values of the hybrids containing PET did not increase with respect to the pure PP. The low viscosity of PET, together with the encapsulation of the clay platelets by this polymer, could be the responsible for this apparent contradiction.

After thermal analysis by DSC (Table 4) the clay was found to be a nucleating agent for the PP crystallinity, as the increase in the crystallization peak temperature revealed. PP/MMT hybrids showed higher melting temperatures than pure PP, indicating higher ordered structure in the PP crystal. No differences in the crystallinity fraction were found in these hybrids. Hybrids containing PET displayed the highest crystallization and melting temperatures and the lowest crystallinity fraction, especially with organophilic clay and MA-g-PP compatibilization. PET was also found to be a nucleating agent for PP crystallinity. Its molecules could not only interact with OMMT clay, but also with PP, through the MA-g-PP functionality. These interactions could be the main cause of reduction of the PP crystallinity.

The hybrids exhibited higher stiffness and tensile strength but lower strain values than pure PP especially when OMMT was used. The higher Young's modulus exhibited by PMOMMT and the high strain at break of PPMMT were especially relevant. As shown in Table 4, the existence of PET did not improve the mechanical properties of the hybrids.

Conclusions

The modification of a natural bentonite, rich in calcium montmorillonite, by ionic exchange reaction with 11-undecyl-ammonium chloride has been shown to be successful. As a result, affinity between the clay and PP was achieved when the organophilic montmorillonite was employed in the synthesis of PP/MMT hybrids by melt compounding.

The improved morphology, as well as the mechanical (stiffness and tensile strength), thermal and flame retardant properties of the hybrids could be due to the polymer molecules intercalation into the montmorillonite galleries. The effect of adding MA-g-PP did not result in significant changes. Hybrids prepared with a blend PP/PET as matrix resulted in improved morphology, which could be due to the higher affinity between PET molecules and the organophilic clay. Only thermal properties were improved with respect to PP/MMT hybrids.

Acknowledgment

Authors thank the financial support provided by DURSI (Government of Catalonia, Spain) and by Ministerio de Educación y Ciencia (Government of Spain; project MAT2004-01563).

[1] E.P. Giannelis, Advanced Materials **1996**, *8*, 29-35.
[2] A. Oya, in: "*Polymer-Clay Nanocomposites*", T.J. Pinnavaia and G.W. Beall, Eds., Wiley, Chichester 2000, pp. 151-172.
[3] X. Liu and Q. Wu, Polymer **2001,** *42*, 10013-10019.
[4] N. Hasegawa, H. Okamoto, M. Kato, A. Usuki, Journal of Applied Polymer Science **2000,** *78*, 1918-1922.
[5] A.D. Brune and J. Bicerano, Polymer **2002,** *42*, 369-387.
[6] K. Yano, A. Usuki, and A. Okada, Journal of Polymer Science. Part A **1993,** *31*, 2493-2498.
[7] R.A. Vaia and E.P. Giannelis, Macromolecules **1997,** *30*, 8000-8009.
[8] I.J. Chin, T. Thurn-Albrecht, H.C. Kim, T.P. Rusell et al., Polymer **2001,** *42*, 5947-5952.
[9] M. Kawasumi, N. Hasegawa, M. Kato, A. Usuki et al., Macromolecules **1997,** *30*, 6333-6338.
[10] J.M. Gloaguen and J.M. Lefebvre, Polymer **2001,** *42*, 5841-5847.
[11] P.H. Nam, P. Maiti, M. Okamoto, T. Kotaka et al., Polymer **2001,** *42*, 9633-9640.

Rheological and Structural Investigation of Layered Silicate Nanocomposites Based on Polyamide or Polyethylene: Influence of Processing Conditions and Volume Fraction Effects

Pascal Médéric,[1] Tolotrahasina Razafinimaro,[1] Thierry Aubry,[1] Michel Moan,[1] Marie-Hélène Klopffer[2]*

[1] Laboratoire de Rhéologie, Université de Bretagne Occidentale, 6, avenue Victor Le Gorgeu, C.S. 93837, 29238 Brest cedex 3, France
Fax: 33 (0)2 98 01 79 30; E-mail: pascal.mederic@univ-brest.fr
[2] Institut Français du Pétrole, B.P. 311, 1-4 avenue Bois-Préau, 92506 Rueil-Malmaison, France

Summary: Organoclay polymer nanocomposites offer improved material properties at very low filler loadings making them of immediate interest for industrial applications. In this study two matrices are used : a polar polymer, PA12, and a non polar polymer, LLDPE. Many articles focused on the importance of the chemistry used to modify the surface of the clay. Numerous papers studied the influence of the compatibilizing agent necessary to obtain intercalated structure, when the matrix is non polar. This paper demonstrates the importance of the way the clay was melt processed into the thermoplastic matrix. Layered silicate nanocomposites based on PA12 or LLDPE were characterized by X-ray diffraction, transmission electron microscopy and rheological tests. The influence of volume fraction on rheological and structural properties is also investigated.

Keywords: layered silicates; nanocomposites; polyamide; polyethylene; processing conditions; rheology

Introduction

Polymer/layered silicate nanocomposites have become an important class of polymeric composites because they exhibit dramatic enhancements in barrier,[1,2] thermal[3,4] and mechanical[5,6] properties even at very low silicate contents. It is generally acknowledged that these properties are strongly affected by the affinity between matrix and clay,[7] the degree of clay exfoliation, the degree of dispersion and the orientation of clay particles throughout the thermoplastic matrix.[8] The chemical strategy used to modify the surface of the clay and the

 DOI: 10.1002/masy.200550308

melt processing conditions play a key role in achieving the structure and properties of the final product.[9] So the study of the rheological behavior of these systems, correlated with other investigation techniques, such as Electron Microscopy and X-Ray Diffraction, is of great importance.[10-12] This work focuses on the influence of the melt-blending conditions and the particle volume fraction on the structure and melt rheological properties of polymer/layered silicate nanocomposites.

Experimental

The first system was prepared from a commercial polymer, a polyamide 12 (Rilsan® AECHVO from Atofina) and a commercial modified montmorillonite clay, referenced as C30B, supplied by Southern Clay Products (Gonzales, TX). The melting point of this PA12 grade is 178°C. Using thermogravimetric analysis, the weight loss of the polyamide 12 used in this work was shown to be less than 0.5% after 45min at 220°C.[13] Its weight and number average molecular weights were shown to be 37 000g/mol and 20 000g/mol respectively.[13] C30B is a methyl bis-2-hydroxyethyl tallow exchanged montmorillonite clay with a modifier concentration of 90 meq/100g. This treatment is adjusted to ascertain a good dispersion in the polar PA matrix and to make a nanocomposite of intercaled or exfoliated type. Indeed, the hydroxyl groups of the organic modifier are able to interact with amine groups of PA via numerous hydrogen bonds.

The second system was made from a commercial linear low density polyethylene (LLDPE, Flerixene FG20F from Enichem) and a commercial modified montmorillonite clay, referenced as C20A, supplied by Southern Clay Products (Gonzales, TX). This PE grade was already used in previous rheological studies.[14,15] Its melting point is 121°C and its weight and number average molecular weights are 140000g/mol and 37000g/mol respectively. C20A is a dimethyl dihydrogenated tallow ammonium montmorillonite clay with a modifier concentration of 95 meq/100g. Due to the low polarity of the alkyl chains, intercalation of polyolefin macromolecules requires the addition of a compatibilizing agent. Anhydride maleic has already been proved to have a high intercalation level with the clay layers.[16-19] This compatibilizer, denoted C, is a maleic anhydride grafted-PE, provided by Atofina. Its melting point is 124°C and its weight and number average molecular weights are 92500g/mol and 20000g/mol respectively.

Both organoclays (C30B and C20A) are thermally stable at the temperatures of this study.[5,20] Composites have been prepared at solid volume fractions Φ ranging from 0% to 12%, by using an internal mixer. Oscillatory and steady shear measurements were performed using a Rheometrics Dynamic Analyser (RDAII) equipped with parallel disks of 25mm diameter and 2mm spacing. Rheological measurements were performed under a nitrogen blanket and carried out at a temperature of 200°C for PA12-based composites, and 195°C for PE-based composites. Data on dynamic moduli and complex vicosity correspond to the higher strain amplitude (4%) for which the response of all systems is linear viscoelastic. PA12 and C30B/PA12 nanocomposites were dried for 4h at 85°C under vacuum before experiments. The structure and morphology of these filled systems were observed by using Wide-Angle X-Ray Scattering, Scanning Electron Microscopy and Transmission Electron Microscopy techniques. X-Ray diffraction patterns were obtained using a Siemens F X-ray diffractometer with Ni-filtered Cu Kα radiation of wavelength 0.15nm, generated at 35kV and 25mA. Data were recorded in the range of $2\theta = 2\text{-}5°$ using a counting time and a step size of 8s and 0.02° respectively. For Transmission Electron Microscopy observations, ultrathin sections were prepared at –100°C with an Ultracut E ultracryomicrotome (Reichert and Jung) using a diamond knife. Imaging was performed with a Zeiss EM 902 at 80kV. The morphology of C20A/PE system (system without compatibilizer) was observed, using a Scanning Electron Microscopy on cryo-fractured surfaces.

Results and Discussion

A. C30B/PA12 nanocomposites

In the first part of this work, the effect of processing conditions (and more particularly the rotational speed of the blades during mixing) on the formation of C30B/PA12 nanocomposites is studied. Three 2.5% C30B/PA12 composite samples, denoted S_1, S_2 and S_3, have been prepared with a mixing time of 6 minutes, that is the time needed to obtain a constant torque on the blades, at different rotational speeds, respectively 32rpm, 65rpm and 100rpm. Figure 1 shows the diffraction patterns from the organically modified clay, C30B, and the systems S_1, S_2 and S_3. The d_{001} basal reflection peak of C30B appears at $2\theta = 4.5°$, corresponding to an interlayer gallery of 1nm. For S_1, this peak is shifted to $2\theta = 2.7°$, i.e. a d-spacing of 2.3nm, suggesting a partial intercalated structure. Indeed PA12 chains diffuse into the host silicate layer galleries, due to the numerous weak interactions between PA12 and modified clay surfaces, thereby producing an expanded polymer-silicate layer structure. The disappearance of this peak, or the reduction in its intensity indicates a partially exfoliated structure for S_2 and S_3.[21] Increasing rotational speed finally leads to exfoliation : platelets move away from each other via shear-induced diffusion of PA12 chains in the organoclay gallery.[22]

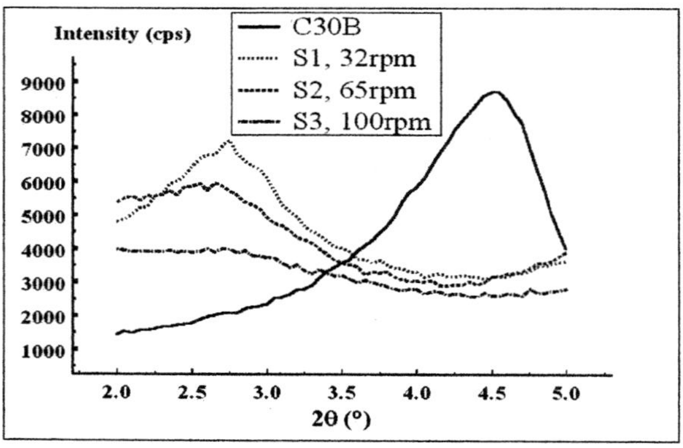

Figure 1. Rotational speed dependence of X-ray diffraction patterns for S_1, S_2 and S_3.

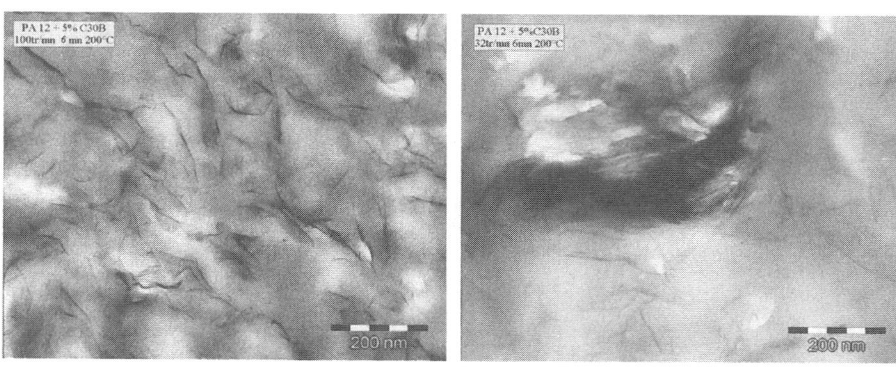

Figure 2. TEM micrographs : a) S_3 system (left), b) S_1 system (right).

Figure 2a shows a TEM micrograph of S_3 system. Exfoliated silicate layers are in a large majority. On the other hand, Figure 2b shows the structure of the S_1 system. In fact, the presence of few exfoliated silicate layers and silicate anisotropic stacks (characteristic size ~500nm) is evidenced in Figure 2b. These observations are in accordance with diffraction spectra of nanocompsite samples.

Figure 3 shows the storage modulus G' versus circular frequency ω, for S_1, S_2 and S_3. S_2 and S_3 nanocomposites display a partial exfoliated nanostructure which leads to both a plateau-like behavior at low frequencies and enhanced viscoelastic moduli (G'' is not presented here) at high frequencies.[19] In addition, the storage modulus increases with increasing rotational speed, especially at the lowest frequencies. The storage modulus level for S_1 is three times lower than that of S_3 in the terminal zone. This result shows that melt processing conditions play a key role in achieving high levels of exfoliation and dispersion. On the other hand the storage modulus values are very similar for S_2 and S_3.

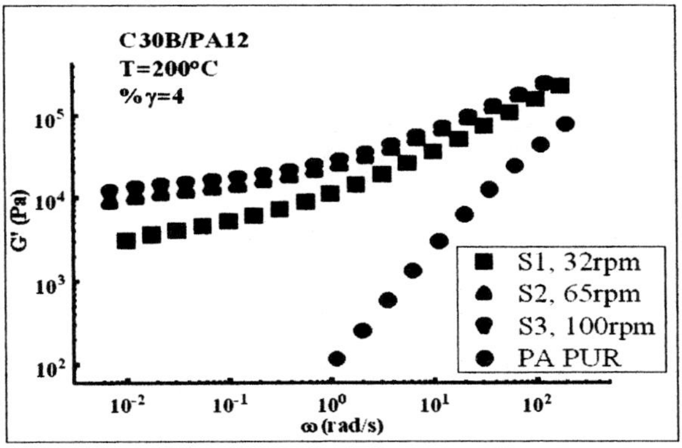

Figure 3. Rotational speed dependence of storage modulus for S_1, S_2 and S_3.

In the second part of this work, the influence of volume fraction on rheological properties of C30B/PA12 nanocomposites is studied. The low frequency relative complex viscosity versus clay volume fraction plot is displayed in Figure 4. Considering the low C30B volume fractions (Φ < 0.75%), the method described by Utracki[23] is used and yields a C30B aspect ratio equal to 140 for the system processed at 32 rpm and 330 for the system processed at 100 rpm. For the last system, assuming a layer thickness of 1nm, the diameter of an isolated clay particle is thus calculated to be 330nm. This result is in very good agreement with average sizes of montmorillonite exfoliated layers.[24] On the other hand, the inferior aspect ratio calculated for the system mixed at 32rpm is indicative of the presence of stacks. These conclusions confirm the XRD and TEM observations for the systems S_3 and S_1.

All results show that optimized processing conditions are a mixing time of 6mn and a rotational speed of 100rpm. Thus, these conditions are used in the following part of this study.

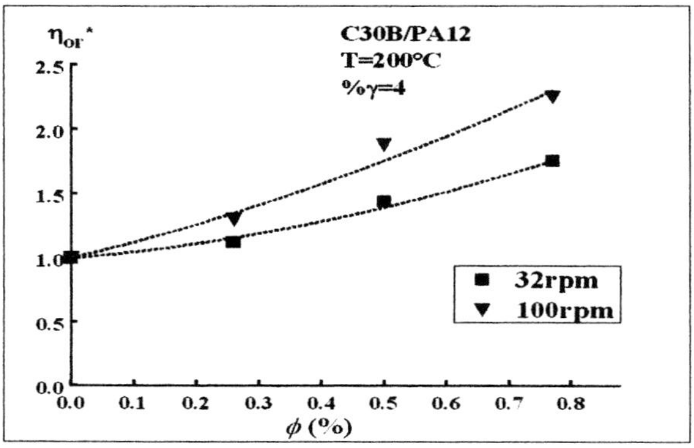

Figure 4. Clay volume fraction dependence of low frequency relative complex viscosity for C30B/PA12.

Figure 5 shows the storage modulus as a function of frequency, in the linear viscoelastic domain for different volume fractions. Although the qualitative behavior of the storage and loss moduli (only G' is presented in Figure 5) are essentially unaffected by increasing clay content in the high frequency range, at low frequencies the viscoelastic behavior changes from liquid-like to solid-like behavior.[25] For sufficiently filled nanocomposites, platelet-platelet interactions are mainly responsible for this change which affects more strongly the storage than the loss modulus. The large enhancement of elastic modulus G' at low frequencies and the frequency independence of the loss tangent G''/G'[25] appear at a critical volume fraction of about 1% and mark the liquid-solid transition. This critical volume fraction is also highlighted in Figure 6 which shows the variation of the first normal stress coefficient

ψ_1 versus shear stress τ. In fact, for volume fractions lower than 1%, a low shear plateau is observed. It is not the case for sufficiently filled systems ($\Phi \geq 1\%$).

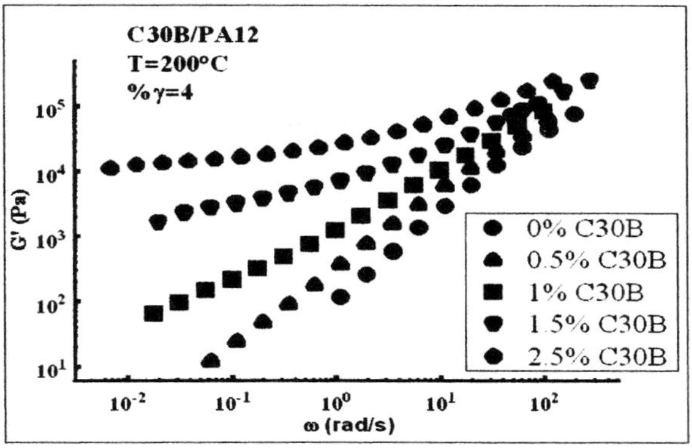

Figure 5. Clay volume fraction dependence of storage modulus for C30B/PA12.

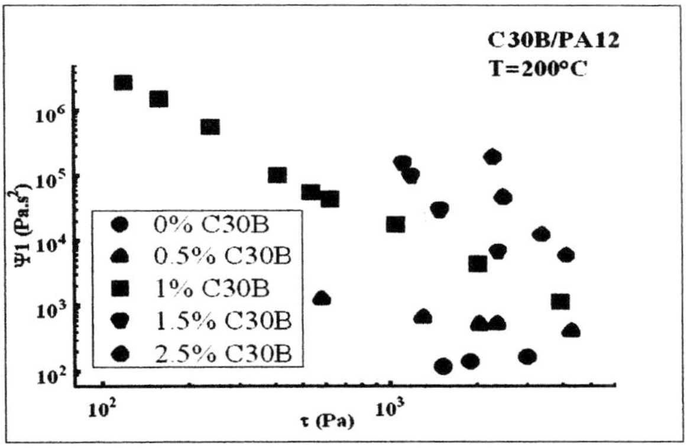

Figure 6. Clay volume fraction dependence of first normal stress coefficient for C30B/PA12.

B. C20A/C/LLDPE Nanocomposite

Because LLDPE does not include any polar group in its backbone, it does not form a nanocomposite structure by simple melt mixing of the organoclay (C20A) and the polymer.

Morphological investigation of C20A/LLDPE systems, as observed by Scanning Electron Microscopy (Figure 7), shows microaggregates formed by primary particles (~ 3μm).
A low frequency response indicative of a solid-like behavior is observed in the G' curve but only for a volume fraction of 12% (Figure 9). Such results are also obtained for conventional composites, such as talc/LLDPE.[26]
2.5%C20A/13%C/LLDPE system was studied. A two-step mixing sequence was used to prepare this system : organoclay/C master batch was prepared by melt mixing first, and then it was melt mixed with LLDPE.
Figure 8 shows the diffraction patterns from the organically modified clay, C20A, and the 2.5%C20A/13%C/LLDPE system. The d_{001} basal reflection peak of C20A appears at 2θ = 3.5°, corresponding to an interlayer gallery of 1.5nm. For 2.5%C20A/13%C/LLDPE system, this peak disappears. This is the signature of an exfoliated structure.
In Figure 9, this nanocomposite shows higher storage modulus over the whole frequency range than the 2.5%C20A/LLDPE. This rheological behavior is in agreement with XRD results. Additional studies are but needed to understand the relation between the structure and the melt rheological properties of C20A/C/LLDPE nanocomposites.

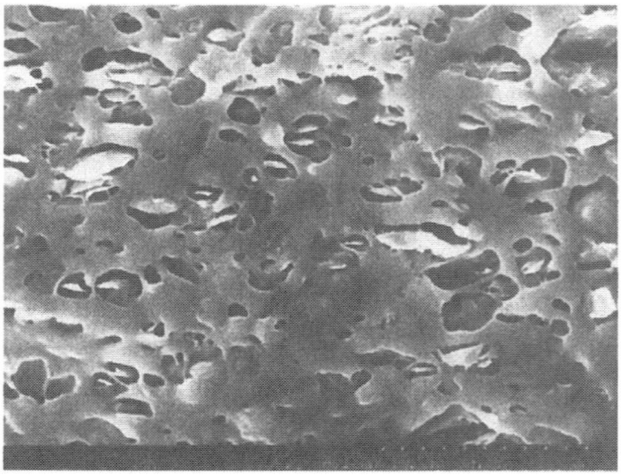

Figure 7. SEM micrographs for 2.5% C20A/LLDPE.

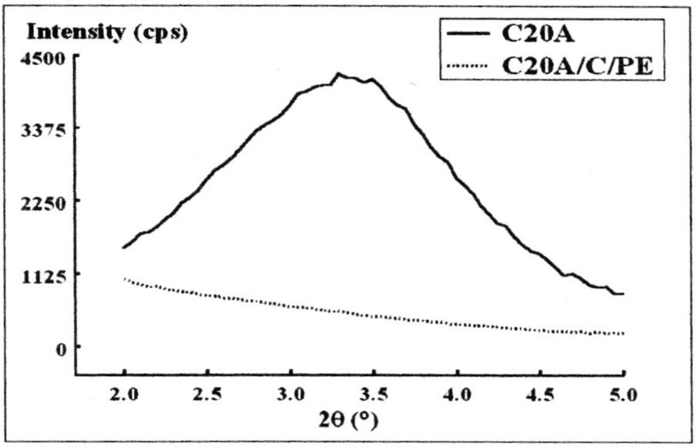

Figure 8. X-ray diffraction patterns for C20A clay and 2.5%C20A/13%C/LLDPE system.

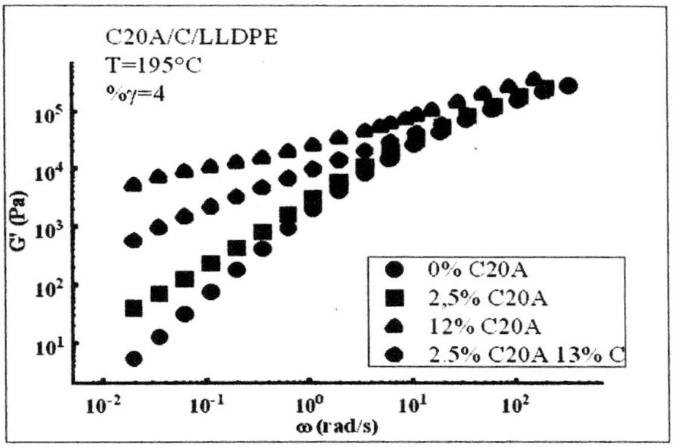

Figure 9. Frequency dependence of storage modulus for C20A/C/LLDPE.

Conclusions

It has been shown that the melt processing conditions have an important influence on the degree of delamination and dispersion of the platelets, thus on the microstructure and properties of the nanocomposite. The relation between the rheological behavior and the structure of C30B/PA12 was discussed : exfoliated structure is revealed not only through a plateau-like G' behavior at low frequency but also enhanced viscoelastic moduli at high frequencies. The influence of the organoclay volume fraction on rheological behavior was studied. Preliminary rheological results concerning C20A/C/LLDPE systems were also presented.

Acknowledgments

The authors express thanks to Dr. F. Michaud, Université de Bretagne Occidentale, and Dr. Y. Fouquet, Ifremer, for XRD tests and Dr. B. Pees, Atofina, for TEM experiments.

[1] M. Krook, A.-C. Albertsson, U. W. Gedde, M. S. Hedenqvist, *Polymer Eng. & Sci.*, **2004**, *42(6)*, 1238.
[2] R. Waché, M.-H. Klopffer, S. Gonzales, P. Médéric, M. Moan, Eurofillers, Alicante, Spain, **2003**, 101.
[3] Y. Tang, Y. Hu, S. Wang, Z. Gui, Z. Chen, W. Fan, *Polymer international.*, **2003**, *52(8)*, 1396.
[4] Y. Tang, Y. Hu, S. Wang, Z. Gui, Z. Chen, *Polymers for adv. Technol.*, **2003**, *14(10)*, 733.
[5] J. W. Cho, D. R. Paul, *Polymer*, **2001**, *42(3)*, 1083.
[6] P. Reichert, J. Kressler, R. Thoann, R. Mülhaupt, G. Stöppelmann, *Acta Polym.*, **1998**, 49, 116.
[7] R. A. Vaia, E. P. Giannelis, *Macromolecules*, **1997**, *30*, 8000.
[8] H. R. Dennis, D. L. Hunter, D. Chang, S. Kim, J. L. White, J. W. Cho, D. R. Paul, *Polymer*, **2001**, *42*, 9513.
[9] S. S. Ray, M. Okamoto, *Prog. Polym. Sci.*, **2003**, *28*, 1539.
[10] J. Ren, A.S. Silva, R. Krishnamoorti, *Macromol.*, **2000**, *33*, 3739.
[11] R. Krishnamoorti, E. p. Giannelis, *Macromol.*, **1997**, *30*, 4097.
[12] D. F. Eckel, M. P. Balogh, P. D. Fasulo, W. R. Rodgers, *J. Applied Polym. Sci.*, **2004**, *93*, 1110.
[13] O. Talon, PhD thesis, INSA Rouen, France, **2002**.
[14] J. Huitric, P. Médéric, M. Moan, J. Jarrin, *Polymer*, **1998**, *39(20)*, 4849.
[15] M. Moan, J. Huitric, P. Médéric, J. Jarrin, *Journal of rheology*, **2000**, *44(6)*, 1227.
[16] A. Usuki, M. Kato, A. Okada, T. Kurauchi, *J. Applied Polym. Sci.*, **1997**, *63*, 137.
[17] M. Kawasumi, N. Hasegawa, M. Kato, A. Okada, A. Usuki, *Macromol.*, **1997**, *30(20)*, 6333.
[18] K. H. Wang, M. H. Choi, C. S. Koo, Y. S. Choi, I. J. Chung, *Polymer*, **2001**, *42*, 9819.
[19] Y. T. Lim, O. O. Park, *Rheologica Acta*, **2001**, *40*, 220.
[20] W. Xie, Z. Gao, K. Liu, W.-P. Pan, R. Vaia, D. Hunter, A. Singh, *Thermochimica Acta*, **2001**, *367/368*, 339.
[21] H. Ishida, S. Campbell, J. Blackwell, *Chem. Mater.*, **2000**, *12*, 1260.
[22] T.D. Fornes, P. J. Yoon, H. Keskkula, D.R. Paul, *Polymer*, **2001**, *42*, 9929.
[23] L. A. Utracki, J. Lyngaae-Jorgensen, *Rheol. Acta*, **2002**, *41*, 394.
[24] G. Lagaly, S. Ziesmer, *Adv. Colloid Interface Sci.*, **2003**,*100/102*, 105.
[25] H. H. Winter, M. M. Mours, *Adv. Polym. Sci.*, **1997**, *134*, 165.
[26] P. Médéric, M. Moan, M.-H. Klopffer, Y. Saint-Gérard, *Applied Rheology*, **2003**, *13(6)*, 297.

Structure/Property Relationships for Polyamide 6 / Organoclay Nanocomposites in the Melt and in the Solid State

Vitaly M. Karaman,[1] *Valery P. Privalko,*[1] *Eleonora G. Privalko,**[1] *Bjorn Lehmann,*[2] *Klaus Friedrich*[2]

[1]Institute of Macromolecular Chemistry, National Academy of Sciences of Ukraine, 02160 Kyiv, Ukraine
E-mail: privalko@iptelecom.net.ua
[2]Institute for Composite Materials (IVW) Ltd, University of Kaiserslautern, D-67663 Kaiserslautern, Germany

Summary: Commercial nanocomposites of polyamide 6 prepared by melt compounding with organoclays were characterized by complex viscosities and relaxation time spectra derived from storage and loss shear moduli measured in the melt state at 230° C, and by wide- and small-angle X-ray scattering, differential scanning calorimetry and stretching calorimetry in the solid state.
In the melt state, the decrease of ~ 25 % (compared to the pristine sample) in Newtonian viscosity at the lowest clay loading (2.5 %) suggested a lower equilibrium elasticity modulus of an entangled melt, as the small amounts of organoclay nanoparticles acted as specific "diluents" for the initial entanglement network. However, by increasing the clay loading this effect disappeared due to the importance of strong interactions at the nanoparticle/melt interface, leading to the formation of a fairly thick boundary interphase (BI) around the nanoparticles and ending up in the build-up of an "infinite cluster" of clay nanoparticles coated with BI at the highest (albeit still unusually low) clay loading (7.5 %).
In the solid state, organoclay nanoparticles proved to induce the crystallographic $\alpha \rightarrow \gamma$ transformation of PA6 , while the matrix crystallinity in nanocomposites remained essentially unchanged. In the range of elastic (reversible) behavior below the apparent yield strains ϵ^* , the highest Young´s moduli E and the lowest linear thermal expansion coefficients α_L were observed for dried nanocomposites, while the lowest E and the highest α_L corresponded to the moisturized, pristine PA6. The endothermal process of shape distortion of the lamellar crystals was assumed to precede the onset of the exothermal process of lamellar fragmentation in the range of inelastic (irreversible) behaviour of PA6 above ϵ^* . The energy balance of the inelastic behaviour of nanocomposites was dominated by the endothermal process of lamellar shape distortion.

Keywords: boundary interphase; clay; nanocomposite; polyamide 6

 DOI: 10.1002/masy.200550309

Introduction

Idealized classical models assuming random distribution of isolated, spherical filler particles within a continuous, structureless polymer matrix [1] , invariably treated the filler volume fraction φ as the major factor controlling the properties of filled polymer composites. In the frame of such "pragmatic" models, the empirical observations of significant increase of, say, elasticity moduli and concomitant decrease of thermal expansivities of filled polymers, the smaller the particle size at the same filler loading, remained unexplained. Formally, this effect can be accounted for by a more realistic "physical" model which explicitly assumes the smearing-out of the "mathematical" (i.e., vanishingly thin) polymer/filler interface into a "boundary interphase" (BI) of quite appreciable thickness [2] . In this context, the contribution of BI to the properties of a polymer/filler composite will increase, leading to a higher total interfacial area (i.e., the smaller the filler particle size).

Apparently, these and similar observations prompted the development of polymer *nanocomposites* reinforced by relatively small amounts of ultrafine clay particles with at least one nano-size dimension and extremely high aspect ratios[3] . The processability of nanocomposites, however, can present some problems, in view of the extreme sensitivity of melt flow behaviour to the aspect ratio of clay nanoparticles [4] and to the quality of their dispersion (exfoliation)[4-7]. Viscoelastic studies of irradiation-grafted nano-inorganic particle-filled polypropylene composites in the melt state[8,9] revealed the onset of a plastic yield phenomenon for a nanocomposite with the filler volume content as low as 4.68 % . This was regarded as experimental evidence for the shear-resistant, infinite cluster of nanoparticles coated with a polymer boundary interphase (BI) when the ratio of the mean thickness of the polymer interlayer between neighbouring nanoparticles, $<L>$, to the mean radius of gyration of a polymer coil, $<R_g>$, approached the "critical" value, $<L>/<R_g> \approx 1$. Moreover, stretching calorimetry studies of these nanocomposites [10] revealed the considerable excess of Young's moduli concomitant to the deficit of thermal expansivities, limiting strains for elastic behavior and breaking strains compared to reasonable theoretical predictions. These results, combined with the X-ray data, were explained by the oriented, pre-stressed state of tie chains in the interlamellar space of the PP matrix.

In this contribution, similar experimental techniques were used to characterize the structure-property relationships for polyamide 6/organoclay nanocomposites in the melt and in the solid state.

Experimental

Materials

The pristine polyamide 6 (designated by the supplier as 200A) and three nanocomposites (299F, 299E and 299D) prepared by melt compounding of 200A with 2.5, 5.0 and 7.5 % , respectively, of organoclays (organically-treated mica-like clay platelets with very high aspect ratio; specific surface area 700 m^2/g) were supplied by RTP Co. (Winona, MN). Both as-received samples with moisture contents up to 2.5 % (series 1) and those evacuated at 393 K overnight to constant weight (series 2) were studied to check for the effects of humidity.

Techniques

Storage $G'(\omega)$ and loss $G''(\omega)$ shear moduli were measured with the PIRSP-03 rheometer in the range of linear viscoelasticity (amplitudes of harmonic vibrations on the order of 1 %) at 230° C in an angular frequency window spanning about three decades ($\omega = 10^{-2}$ to 10 rad/sec). The poor thermal stability of 200A prohibited measurements at higher temperatures.

Wide-angle X-ray scattering (WAXS) patterns in the range of scattering angles (2θ) from 5 to 40 deg were measured with a DRON-2,0 diffractometer ; small-angle X-ray scattering (SAXS) data in the scattering angles range from 3' to 5 deg were obtained with the standard Kratky camera (KRM-type diffractometer) as described elsewhere [10].

Specific heat capacity c_p was determined (relative error of a single run about 3 %) with a custom-made differential calorimeter with diathermal shells in the temperature interval from 150 to 550 K (that is, up to 50 K above the apparent melting point of 500 K).

The mechanical work (W) and concomitant heat effects (Q) in the step-wise loading (stretching)/unloading (contraction) cycles were measured (with the estimated mean error below 2 %) at room temperature by using the stretching calorimeter described elsewhere [10]. In every experimental run, each specimen was stretched at a constant velocity q^+ (10% of the total specimen length per minute) to a predetermined strain ε_i , stored at fixed ε_i to the full completion of mechanical and thermal relaxations, and thereafter allowed to contract at the same velocity q^- to zero force.

Results and discussion

1. Melt state

Both for the pristine polymer (sample 200A) and for the nanocomposite with the lowest organoclay loading (sample 299F) the loss moduli $G''(\omega)$ were considerably higher than the storage moduli $G'(\omega)$; moreover, both $G''(\omega)$ and $G'(\omega)$ exhibited a smooth increase with frequency ω. In contrast, one could identify at least two frequency intervals with distinctly different behaviour for the nanocomposite with the highest organoclay loading (299D). In the range of lowest frequencies (i.e., at $\omega < 0.1$ sec^{-1}), both $G'(\omega)$ and $G''(\omega)$ turned out to have nearly identical magnitudes and to be only weakly frequency-dependent. At higher frequencies, $G'(\omega)$ and $G''(\omega)$ started to diverge and to exhibit a frequency dependence (although much weaker than that for the pristine polymer). These data may be considered [8,9] as experimental evidence for the existence of a spatial network of filler particles coated with a polymer boundary layer in the low-frequency range, and for its plastic yield at higher frequencies for sample 299D. As might have been expected, the viscoelastic behaviour of the sample 299E is intermediate between those for two other nanocomposites.

In principle, the occurrence of the yield phenomenon in molten polymer composites should be directly observable at exceedingly low frequencies ; otherwise, it can be associated with a non-zero intercept on the square-root plots of $G'(\varphi)$ vs. $G'(0)$, as follows from the modified Casson equation [11]:

$$\sigma(\varphi)^{1/2} = \sigma_y^{1/2} + a\,\sigma(0)^{1/2}\,, \qquad (1)$$

where $\sigma(0)$ and $\sigma(\varphi)$ are the shear stresses for the matrix and for the composite, respectively, σ_y is the yield stress, and a is the fitting constant. As can be seen from Figure 1, for nanocomposites with low filler loading the intercepts are negligibly small, whereas for sample 299D one derives $\sigma_y \approx 30$ Pa.

As could be inferred from the approximate constancy of the complex viscosity $|\eta^*| = [(G'/\omega)^2 + (G''/\omega)^2]^{1/2}$ over fairly broad frequency intervals, in the range of low frequencies the samples 200A and 299F behaved as Newtonian liquids. Unexpectedly, the values of $|\eta^*|$ for the latter sample turned out lower by nearly 25% as if the small amounts of organoclay nanoparticles acted as specific "diluents" for the initial entanglement network. In contrast, no evidence for the Newtonian flow could be detected for the nanocomposite

200D with the highest organoclay loading ; instead, the complex viscosity $|\eta^*|$ strongly increased, more markedly for lower frequency. This behaviour confirmed the occurrence of plastic yield of an infinite cluster of nanoparticles coated with BI, preceding the macroscopic melt flow. The 299E nanocomposite exhibited an intermediate behavior.

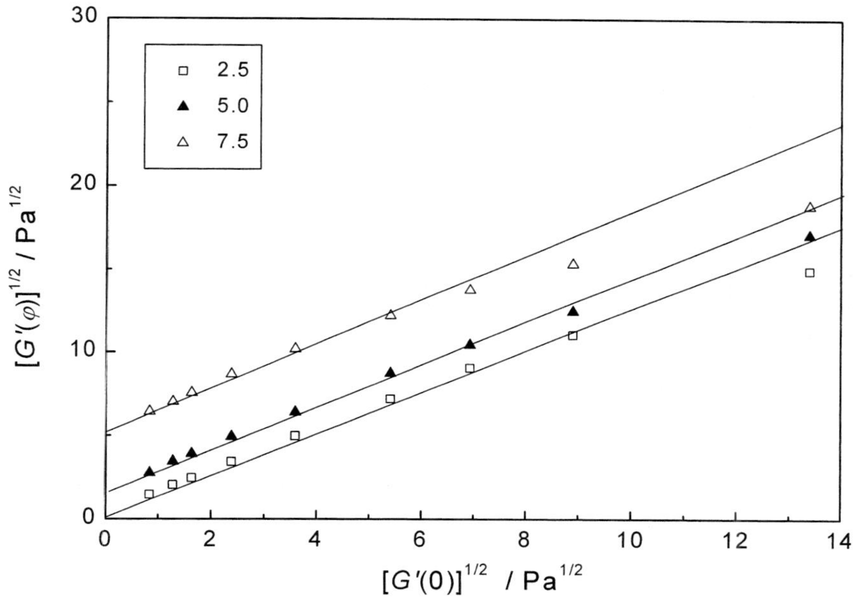

Figure 1. Modified Casson plots for nanocomposites with different organoclay loading.

2. Solid state

Reasonably strong WAXS reflections at the scattering vectors q = 14.4 and 16.4 nm^{-1} for the 200A nanocomposite (Figure 2a) were the evidence for predominance of monoclinic (a) crystal phase with hydrogen bonds between antiparallel chains in the pristine semi-crystalline PA6. These reflections could be also observed on the WAXS patterns of nanocomposites as shoulders flanking the main diffraction maximum at q = 15.0 nm^{-1}. The latter maximum and the subsidiary reflection at q = 7.5 nm^{-1} are characteristic for the pseudo-hexagonal (or monoclinic) crystal unit cell with hydrogen bonds between parallel chains (γ–modification) . The well-resolved SAXS reflection near $q \approx 0.62$ nm^{-1} (corresponding to the Bragg's periodicity $D \approx 10.1$ nm) for the 200A nanocomposite (Figure 2b) could not be resolved on the SAXS curves of nanocomposites due to the increase of SAXS intensities in the same range of scattering vectors. These results suggested a significant increase of structural heterogeneity due to the appearance of new, strongly scattering entities (presumably, polymer-nanoparticle interfaces and microvoids) with a broad distribution of sizes.

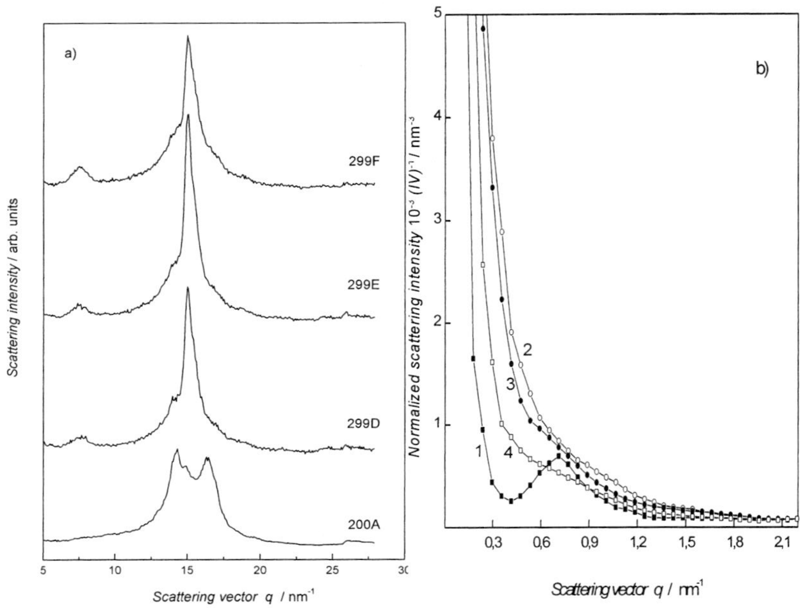

Figure 2. WAXS (a) and SAXS (b) patterns for 200 A (1), 299 D (2), 299 E (3) and 299 F (4).

In the range of elastic (reversible) deformations below a certain characteristic strain ε^* both the specific (per sample mass m) mechanical work (W/m) and specific heat effects (Q/m) data for all studied samples could be quantitatively fitted (Figure 3) to the classical equations of the thermoelasticity of solids [12]

$$W/m = E\varepsilon^2/2\rho \qquad (2a)$$

$$Q/m = E\alpha_L T\varepsilon/\rho \qquad (2b)$$

where E is the Young´s modulus, α_L is the linear thermal expansion and ρ is the density. As expected, the best-fit values of E and α_L for moisture-plasticized samples of series 1 were systematically lower and higher, respectively, than those for corresponding dried samples of series 2. The observed overshoots of E, as well as the deficit of α_L for nanocomposites (Table

1) compared to theoretical predictions were regarded as experimental evidence for the oriented state of tie-chains in the interlamellar space of the PA matrix.

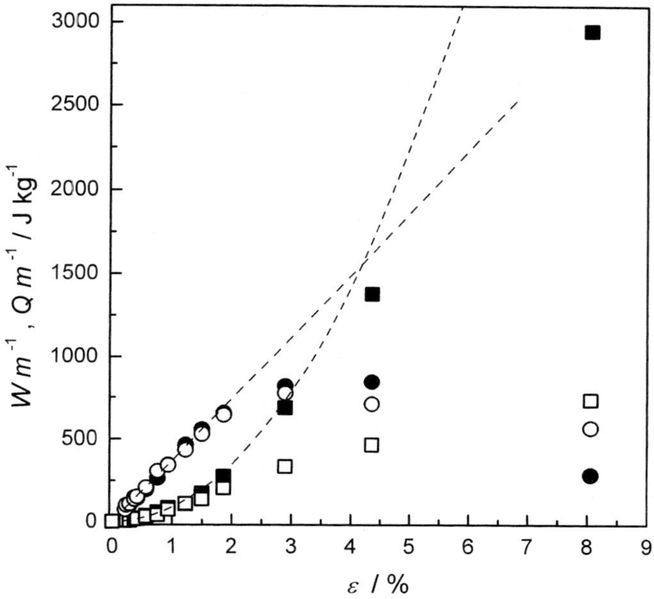

Figure 3. Specifi Figure 4. Specific internal energy increments for nanocomposites. Dashed line: theoretical curve for the hypothetical elastic behaviour of sample 200 A.

Table 1. Crystallinity (WAXS) and thermoelastic parameters for samples of series 1 and 2.

Sample	Clay content / %	X_{WAXS}	E / GPa	$10^4\,\alpha_L$ / K^{-1}	ε^* / %	ε' / %	ε_b / %
200A	0	0.28/0.28	1.0/2.0	0.86/0.73	1.5/1.7	2.2/2.3	93/14.9
299D	7.5	0.34/0.38	1.6/2.7	0.80/0.60	1.3/1.4	2.2/2.0	11/3.9
299E	5.0	0.37/0.39	1.5/2.6	0.82/0.66	1.5/1.7	2.1/2.2	13/3.6
299F	2.5	0.30/0.29	1.3/2.4	0.87/0.71	1.7/1.8	2.3/2.2	23/1.8

The increasing differences between the values of W/m in stretching/contraction cycles above the apparent yield strains ε^* (Figure 3) are the evidence for the onset and subsequent development of irreversible (inelastic) structural changes within the polymer matrix (commonly referred to as plastic flow). Importantly, the deviations of experimental values of both W/m and Q/m in stretching regime from theoretical curves (dashed lines) set on at considerably higher strains ε' compared to ε^*. These data suggest the dominant contribution of a certain endothermal phenomenon into the mechanism of plastic flow of PA 6 in the strain interval $\varepsilon^* < \varepsilon < \varepsilon'$, while the contribution of the subsequent exothermal process responsible for the apparent deficit of Q/m becomes significant only above ε'. At this moment, the detailed mechanisms of the observed endothermal and exothermal phenomena remain unclear ; presumably, these may be attributed to successive processes of *shape distortion of the lamellar crystals* and of *lamellar fragmentation*, respectively [12]. In this context, the initial overshoot of W/m in stretching above that in contraction may be regarded as a measure of the specific mechanical energy irreversibly spent in the process of lamellar shape distortion.

The irreversible phenomena proved to set on at about the same apparent yield strains ε^* for the pristine polymer and for the nanocomposites. Formally, these data imply negligible (if any) contribution of *debonding* at the nanoparticle-matrix interface to the possible mechanisms of inelastic deformations. Additional check of this hypothesis can provide the analysis of specific internal energy increments, $\Delta U/m = (W + Q)/m$. As argued elsewhere [13], the deficit of $\Delta U/m$ in filled samples compared to that for the neat matrix polymer ($\delta(\Delta U/m) < 0$) would be considered as evidence for the heat released due to the formation of a free (debonded) filler surface. However, in contrast to the case of polyethylene/ kaolin composites [7] one observes the significant excess of $\Delta U/m$ with respect to the pristine PA6 (that is, $\delta(\Delta U/m) > 0$) for all nanocomposites (Figure 4). It can be thus concluded that of all possible mechanisms of irreversible structural rearrangements above ε^*, it is the endothermal process of lamellar shape distortion which dominates the energy balance of inelastic behaviour of nanocomposites.

Finally, smaller values of the breaking strains ε_b for nanocomposites compared to the pristine sample (Table 1) were attributed to the increased density of fibrillation and cavitation sites (presumably, localized at polymer/nanoparticle interfaces).

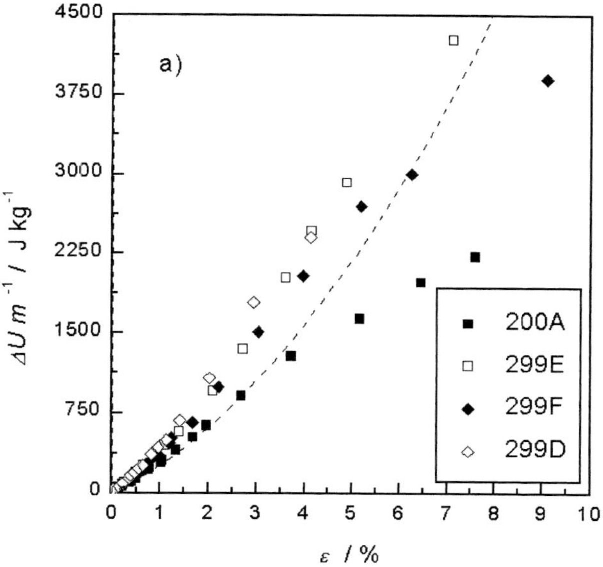

Figure 4. Specific internal energy increments for nanocomposites. Dashed line: theoretical curve for the hypothetical elastic behaviour of sample 200 A.

Conclusions

1. Strong interactions at the nanoparticle/melt interface are responsible for the formation of a fairly thick boundary interphase (BI) around the nanoparticles ending up in the build-up of an "infinite cluster" of clay nanoparticles coated with BI at the highest (albeit still unusually low) clay loading (7.5 %).

2. Organoclay nanoparticles induce crystallographic $\alpha \rightarrow \gamma$ transformation of PA6 , while the matrix crystallinity in nanocomposites remains essentially unchanged.

3. The endothermal process of shape distortion of the lamellar crystals is assumed to precede the onset of the exothermal process of lamellar fragmentation in the range of inelastic (irreversible) behaviour of PA6 above ε^* .

4. The energy balance of the inelastic behaviour of nanocomposites is dominated by the endothermal process of lamellar shape distortion

[1]. L.E. Nielsen, *"Mechanical Properties of Polymers and Composites"*, Marcel Dekker, Inc., New York 1974.
[2]. V.P. Privalko, V.V. Novikov, *"The Science of Heterogeneous Polymers"*, J. Wiley & Sons, Chichester, 1995.
[3]. Y. Fukushima, S. Inagaki, *J. Inclusion Phenom.* **1987**, *5*, 473-482.
[4]. K.H. Wang, M. Xu, Y.S. Choi, I.J. Chung, *Polymer Bull.* **2001**, *46*, 499-505.
[5]. E.P. Giannelis, *Adv. Mater.* **1996**, *8*, 29-35.
[6]. B. Hoffmann, J. Kressler, G. Stoeppelmann, C. Friedrich, G.M. Kim, *Colloid & Polymer Sci.* **2000**, *278*, 629-636.
[7]. Y.H. Hyun, S.T. Lim, H.J. Choi, M.S. Jhon, *Macromolecules* **2001**, *34*, 8084-8093.
[8]. V.F. Shumsky, E.G. Privalko, V.M. Karaman, V.P. Privalko, R. Walter, K. Friedrich, M.Q. Zhang, M.Z. Rong, *Adv. Compos. Let.* **2001**, *10*, 181-185.
[9]. V.P. Privalko, V.F. Shumsky, E.G. Privalko, V.M. Karaman, R. Walter, K. Friedrich, M.Q. Zhang, M.Z. Rong, *Sci. Technol. Adv. Mater.* **2002**, *3*,111-116.
[10]. V.P. Privalko, V.M. Karaman, E.G. Privalko, R. Walter, K. Friedrich, M.Q. Zhang, M.Z. Rong, *J. Macromol Sci. Phys.* **2002**, *41B*, 485-503.
[11]. M.M. Dumoulin, L.A.Utracki, P.J. Carreau, in: *"Two-Phase Polymer Systems"*, L.A. Utracki, Ed., Hanser Publ., Munich 1991, p. 185-212.
[12]. R.J. Young, *"Introduction to Polymers"*, Chapman and Hall 1989.
[13]. V.P. Privalko, F.J. Balta-Calleja, D.I. Sukhorukov, E.G. Privalko, R. Walter, K. Friedrich, *J. Mater. Sci.* **1999**, *34*, 497-508.

Macromol. Symp. **2005**, *221*, 95-104

Starch Nanocrystal Fillers in an Acrylic Polymer Matrix

*Hélène Angellier,**[1] *Jean-Luc Putaux,*[1] *Sonia Molina-Boisseau,*[1] *Danielle Dupeyre,*[1] *Alain Dufresne*[2]

[1] Centre de Recherches sur les Macromolécules Végétales (CERMAV-CNRS), Université J. Fourier, BP 53, 38041 Grenoble Cedex 9, France
E-mail: angellier@cermav.cnrs.fr
[2] Ecole Française de Papeterie et des Industries Graphiques de Grenoble (EFPG-INPG), BP 65, 38402 St Martin d'Hères Cedex, France

Summary: Aqueous suspensions of nanocrystals can be obtained by acid hydrolysis of native waxy maize starch granules. The disruption of waxy maize starch granules by 2.2N HCl hydrolysis has been followed by scanning and transmission electron microscopy and laser granulometry. The mechanical properties of composite materials made of poly(styrene-co-butyl acrylate) filled with starch nanocrystals were characterized by dynamic mechanical analysis. These nanocrystals appeared to be an interesting reinforcing phase in a thermoplastic matrix.

Keywords: acid hydrolysis; filler; nanocomposites; starch nanocrystals; waxy maize starch

Introduction

The growing interest in degradable products has led scientists to develop the use of biopolymers for applications in which synthetic polymers or mineral fillers were traditionally used. In this context, research has been devoted to cellulose and especially to cellulose which revealed to be an interesting model filler in various polymer matrices [1]. By analogy with cellulose whiskers, studies have been performed on starch. Starch is a natural and biodegradable polysaccharide presents in higher plants as an energy storage polymer. Starch granules are mainly composed of two polysaccharides consisting of α-linked D-glucosyl units: amylose (a mostly linear macromolecule) and amylopectin (a 4-5% branched macromolecule). The granules exhibit an onion-like structure of alternating amorphous and semi-crystalline growth rings. It is accepted that the partial crystallinity of native starch granules is due to a clustered organization of

DOI: 10.1002/masy.200550310

amylopectin side chains [2]. Aqueous suspensions of nanocrystals can be obtained after several weeks of hydrochloric acid hydrolysis (also called "lintnerization") of native starch granules [2,3]. In a previous work, semi-crystalline residues obtained by this treatment on potato starch appeared to be an interesting reinforcing phase in poly(styrene-co-butyl acrylate) [4]. As it has been recently shown [5], crystalline nanoplatelets with a well-defined morphology can be obtained from the lintnerization of amylopectin-rich "waxy" maize native starch granules. The present work focuses on studying the evolution of starch granules morphology during acid treatment as well as the influence of selected parameters of the acid hydrolysis (nature and concentration of acid, time of hydrolysis, temperature) on the disruption of such starch granules. The insoluble residues were characterized by scanning and transmission electron microscopy (SEM and TEM), as well as laser granulometry. The mechanical properties of composite materials made of poly(styrene-co-butyl acrylate) (poly(S-co-BuA)) filled with starch nanocrystals were also analyzed by dynamic mechanical analysis (DMA).

Experimental

Waxy maize starch nanocrystals. 10 g of native amylopectin-rich (99 wt%) waxy maize starch granules (Waxylis™, Roquette S.A., Lestrem, France) was mixed with 200 mL of diluted acid (2.2N HCl or 1.5-4M H_2SO_4) in order to reach a weight concentration of 5 wt% [3-6]. Then, the suspensions were put in a thermostated atmosphere (35°C or 40°C) and continuously stirred at 100 rpm. After different times of hydrolysis, the insoluble residues were washed by successive centrifugations in distilled water until neutrality and mechanically treated by Ultra Turrax (IKA, Staufen, Germany) during 5 min at 13000 rpm in order to disperse the aggregates formed during centrifugation. The resulting aqueous suspensions were stored at 4°C. The yield of hydrolysis Y was calculated as the ratio of the dry weight of insoluble residue after hydrolysis to the dry weight of native granules.

Poly(styrene-co-butyl acrylate). We used a latex of a copolymer of styrene (34 wt%) and butyl acrylate (64 wt%) stabilised by 1% acrylic acid and 1% acrylamide, with a glass-rubber transition temperature (Tg) around 0°C. The latex contained particles with an average diameter of 200 nm.

Composite film processing. The starch nanocrystals suspensions treated by Ultra Turrax were mixed with the polymer latex in fractions varying from 0 to 50 wt%. The mixtures were freeze-dried, heated at 90°C during 10 min and hot-pressed during 1 min at 2 MPa (300 psi) to obtain films about 1 mm thick.

Electron microscopy. Scanning electron microscopy was performed with a JEOL JSM-6100 (Tokyo, Japan) instrument. Transmission electron microscopy observations were made with a Philips CM200 (Eindhoven, The Netherlands) microscope operated at 80 kV. Particles were negatively stained with 2 %(w/v) uranyl acetate.

Laser granulometry. Laser granulometry measurements were done using a Malvern Mastersizer (Orsay, France). The dispersions were characterized using the median particle size d_{50}, which divides the size distribution into two equal halves.

Dynamic mechanical analysis. Dynamic mechanical tests were carried out using a RSA2 spectrometer from Rheometrics (Piscataway, New Jersey, USA) working in the tensile mode. The samples were thin rectangular strips with dimensions about $25 \times 7 \times 0.7$ mm^3. The setup measured the complex tensile modulus E*, that is displayed as the storage component E' and the loss component E". Measurements were performed in isochronal conditions at 1 Hz and the temperature was varied between 200 and 400 K by steps of 3 K with a soak time of 30 s.

Results and discussion

Hydrolysis kinetics. First, hydrolysis was performed using the conditions of lintnerization defined by Robin *et al.* [3]: 2.2N HCl at 35°C, during 2 to 40 days. When Y is plotted versus time, three phases are distinguished (Figure 1a) corresponding to different stages of granule disruption. During the first 4 days, the yield rapidly drops to 38 wt%, which roughly corresponds to the relative crystallinity of the native granules [7]. Considering that α(1-6) links are more reactive than α(1-4) and that acid diffusion is favoured in disorganized regions, the amorphous growth rings are rapidly dissolved during this first stage. From 4 to 14 days, the yield decreases more slowly to 10 wt%.

During this stage, the acid gets enough time to diffuse in the semi-crystalline growth rings and preferentially dissolves the α(1-6) branching points. The yield finally reaches 0.5 wt% during the remaining 26 days. Crystallites containing mostly α(1-4) links and some residual α(1-6) links are slowly hydrolyzed.

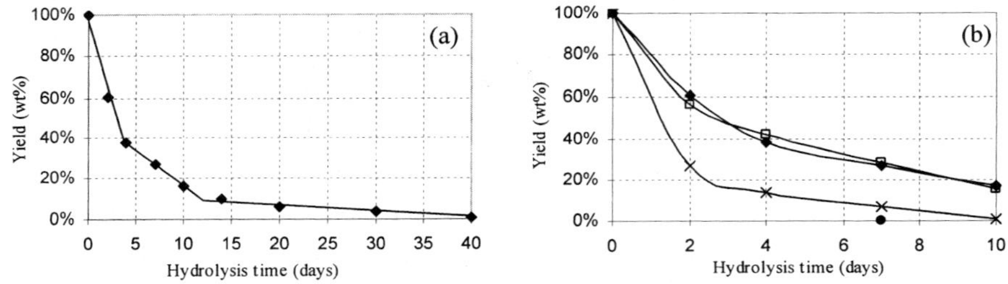

Figure 1. Evolution of the yield of hydrolysis Y versus time of (a) 2.2N HCl at 35°C, (b) 2.2N HCl at 35°C (♦), 3M H_2SO_4 at 35°C (□) and 40°C (x).

We have restricted our examination of the effect of others parameters on the H_2SO_4 hydrolysis to the first stages of acid treatment (2, 4, 7 and 10 days for the nature of acid and temperature, 7 days for the acid concentration). By comparison with 2.2N HCl treatment, a 3M H_2SO_4 hydrolysis is needed to obtain the same evolution of yield (Figure 1b) as already shown by Muhr *et al.* [8]. In addition, a higher temperature (40°C) accelerates the disruption of the granules (Figure 1b). Like for temperature, an increase in acid concentration dramatically accelerates the degradation. Yields of 43.7, 28.6 and 0 wt% were respectively obtained for 1.5M, 3M and 4M H_2SO_4 at 35°C during 7 days. A 4M concentration leads to a complete dissolution of starch.

Evolution of the insoluble residue morphology. The major fraction of native waxy maize starch consists of 10-15 μm polyhedral granules (Figure 2a). After 2 days of HCl treatment, the granule surface is rougher. As the acid hydrolysis progresses, pits develop, which may correspond to erosion or a collapse of the faces (Figures 2a-c). This phenomenon is faster with H_2SO_4 (not shown). Figure 2c suggests that the edges are more resistant than the faces. The initial morphology of granules is lost after 10 days. Aliquots of native and hydrolyzed granules were manually sheared in conical tubes. Broken native granules present a radial fibril-like fracture surface (Figure 2d). This structure has already been described for starches from other

Figure 2. SEM pictures of native waxy maize starch granules (a,d) and residues obtained after 2 days (e), 4 days (b,f) and 7 days (c) of 2.2N HCl hydrolysis. Some samples were broken by manual shearing in a conical tube before (d) or after hydrolysis (e,f).

botanical origins [9-15]. Irregular cavities can be observed at the hilum of the granules (Figure 2d). When the granules are dried, cracks may develop at the central region which is assumed to be less organized than the rest of the particle [16]. After 2 days of hydrolysis, both radial fibril-like organization and 250-400 nm thick concentric layers are observed (Figures 2e,f). This illustrates the onion-like structure of alternating amorphous and semi-crystalline layers also shown by TEM [17]. After 4 days, only fragments of the outer growth rings are observed, confirming that they are more resistant than the inner region. These results are consistent with a radial diffusion of the acid to the less resistant granule core. This diffusion may be favoured by the existence of pores running from the surface to the hilum. This degradation mechanism is comparable to that of α-amylase enzymatic hydrolysis of potato starch granules [18,19].

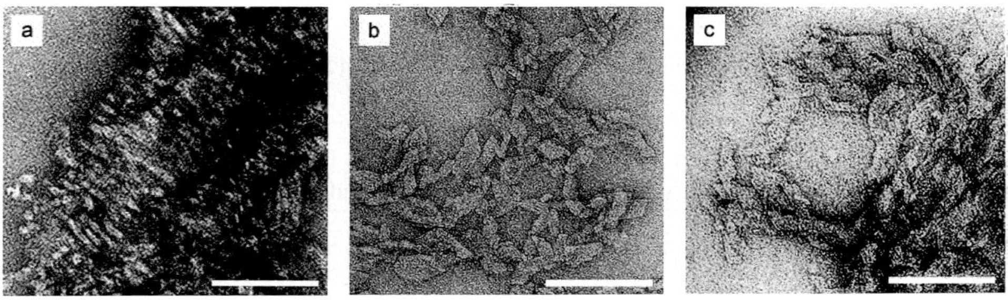

Figure 3. TEM pictures of negatively stained nanocrystals after (a) 4 days and (b) 14 days of 2.2N HCl hydrolysis, (c) 14 days of 3M H_2SO_4 hydrolysis. Scale bar: 100 nm.

After a few days of 2.2N HCl hydrolysis, TEM pictures show fragments containing ordered stacks of lamellae oriented edge-on (Figure 3a). They are assumed to correspond to the organization of amylopectin in the semi-crystalline growth rings: amorphous lamellae containing α(1-6) branching points alternating with crystalline lamellae consisting of double-helices of amylopectin side-chains [5,10,11]. The thickness of the crystalline lamellae measured from the pictures is about 6-7 nm and the combined thickness of crystalline and amorphous lamellae is about 9-10 nm. These data are consistent with those from previous works [10,11,17,20,21]. Such ordered fragments are seen during the first 10 days of hydrolysis. This shows that in the first phase of hydrolysis, the semi-crystalline growth rings are also partially susceptible to hydrolysis. From 7 days of hydrolysis, some parallelepiped platelets (20 nm × 50 nm) are observed, which correspond to separated lamellae seen in planar view [5]. Typical images of such platelets obtained after 14 days of hydrolysis are shown Figure 3b. Aggregates of platelets with similar shapes were also observed in a sample treated for 14 days with 3M H_2SO_4 (Figure 3c). The fact that the yield of hydrolysis rapidly gets lower than the relative crystallinity of native starch granules (about 40% [7]) and that it drops to 0.5 wt% after 40 days of hydrolysis suggests that the crystalline platelets are dissolved by the acid after being separated.

Table 1. Evolution of the median size d_{50} of the insoluble residue during 2.2N HCl and 3M H_2SO_4 hydrolysis for 5 min of Ultra Turrax treatment.

Time of hydrolysis (days)	2	7	10	14	20	40
d_{50} (μm), 2.2N HCl	14.4	11.4	8.3	8.0	7.2	6.3
d_{50} (μm), 3M H_2SO_4	13.7	8.0	5.6	5.3	5.6	-

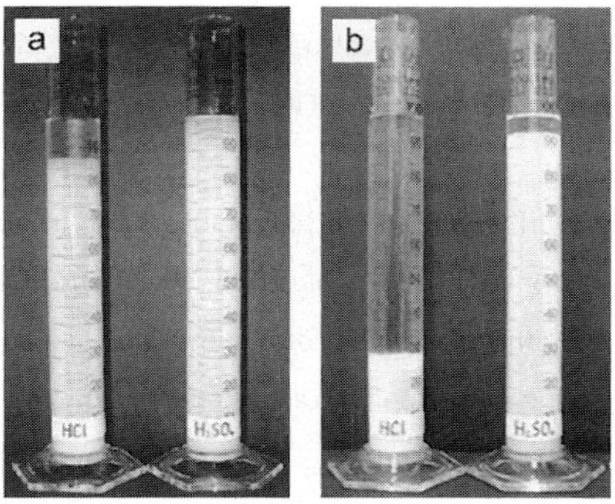

Figure 4. Comparison of the sedimentation properties of HCl (left tube) and H_2SO_4 (right tube) hydrolysis residues suspended in water: (a) after 5 min ; (b) after 60 min.

Table 1 shows the regular decrease of the median particle size d_{50} measured by laser granulometry in 0.02 wt% suspensions after different times of 2.2N HCl or 3M H_2SO_4 treatment. This decrease is consistent with the disruption of the granules observed by microscopy techniques. However, the sizes measured after 2 weeks of hydrolysis (8.0 μm for HCl and 5.3 μm for H_2SO_4) are in apparent contradiction with the dimensions of the nanometric platelets observed by TEM (Figures 3b,c). In fact, both techniques bring complementary data on the same sample but at different scales. Laser granulometry clearly detects micrometric objects that are composed of the nanometric units shown by TEM. That d_{50} does not vary much after 14 days of acid treatment which suggest that flocculation occurs in the suspensions. While the hydrolysis severs an increasing number of α(1-6) branching points, the platelets that should be more separated tend to re-agglomerate by electrostatic interaction. The formation of sulfate-ester groups on the surface of the nanocrystals during a H_2SO_4 hydrolysis should limit this flocculation and produce more stable suspensions. This was confirmed by comparing the sedimentation properties of suspensions obtained using each acid (Figure 4). One other possible reason for improved stability in the case of H_2SO_4 hydrolysis may be due to the lower particle.

Mechanical behaviour of composite materials. The evolution of the logarithm of the storage tensile modulus E' as a function of temperature has been studied for poly(styrene-co-butyl acrylate) films reinforced by starch nanocrystals obtained from 2.2N HCl hydrolysis at 35°C. The influence of filler concentration is displayed in Figure 5 after 7 days of hydrolysis. The curve corresponding to the unfilled matrix is typical of an amorphous thermoplastic behavior.

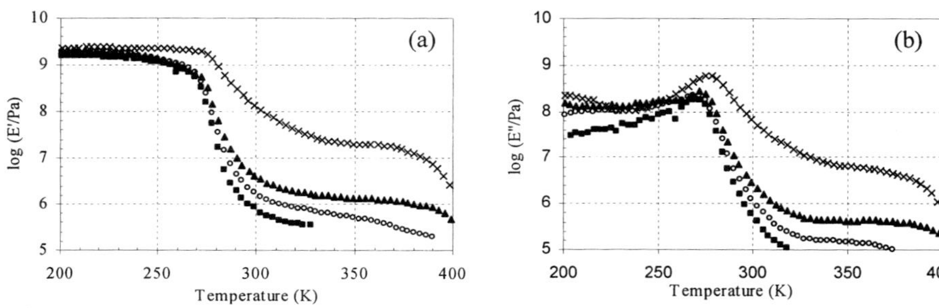

Figure 5. (a) Logarithm of the storage modulus E' and (b) logarithm of the loss modulus E" versus temperature at 1 Hz for composite materials filled with 0 (■), 10 (○), 30 (▲) and 50 wt% (x) of starch nanocrystals obtained from 2.2N HCl hydrolysis at 35°C during 7 days.

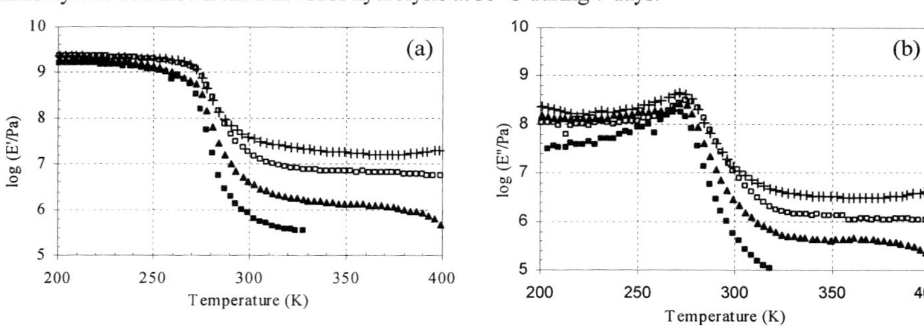

Figure 6. Logarithm of the storage modulus E' (a) and logarithm of the loss modulus E" versus temperature at 1 Hz for composites materials filled with 0 (■) and 30 wt% of starch fillers obtained from 2.2N HCl hydrolysis at 35°C during 7 days (▲), 14 days (□) and 40 days (+).

For temperatures below Tg (273 K), the copolymer is in the glassy state and the storage modulus E' (around 2 GPa) slightly decreases with temperature. Then, E' decreases rapidly by more than 3 decades. This sharp drop is associated with the glass-rubber transition of the polymer. In the terminal zone, E' decreases with temperature until irreversible flow of the material. For temperatures above Tg, the composite modulus of the 50 wt% filled material is about 100 times higher than that of the matrix. In addition, the flowing of the composite films appears at higher temperature that for the unfilled matrix. The plot of log(E') versus temperature for poly(S-co-BuA) reinforced with 30 wt% starch nanocrystals obtained for different times of hydrolysis is displayed in Figure 6a. The storage modulus increases with increasing times of hydrolysis, i.e. with a decrease of the size of the nanocrystals aggregates (Table 1). In fact, for a given starch content, the lower the size of particles, the higher the dispersion of the filler among the matrix, what leads to a higher storage modulus. Figures 5b and 6b show that no shift of the main relaxation (associated with T_g) is observed.

Conclusions

Agglomerates of semi-crystalline starch nanoplatelets were obtained after acid hydrolysis of waxy maize starch granules under variable conditions. Starch nanocrystals obtained after 40 days of 2.2N HCl hydrolysis appear to be an interesting filler in a poly(S-co-BuA) matrix. The mechanical properties of the composite materials are substantially improved by increasing the amount of filler. The problem is that such a filler obtained with a yield of 0.5% and suspensions are not stable. Further experiments (tensile tests, Differential Scanning Calorimetry and Thermo Gravimetric Analysis) have to be carried out in order to investigate the origin of the reinforcing effect of starch filler and the influence of the morphology of nanoplatelets.

Further work will consist in preparing stable suspensions of starch nanocrystals, in shorter time and with a higher yield. We have shown that H_2SO_4 hydrolysis produces more stable suspensions and that an increase of the temperature and/or acid concentration accelerates the degradation of starch granules. A design of experiments is in progress in order to optimize H_2SO_4 hydrolysis, taking into account different selected parameters: hydrolysis time, temperature, acid concentration, starch concentration and stirring.

Acknowledgments

The authors thank Roquette (Lestrem, France) for the supply of waxy maize starch and Dr A. Buléon (LPCM, INRA Nantes, France) for helpful discussions.

[1] V. Favier, G.R. Canova, J.Y. Cavaillé, H. Chanzy, A. Dufresne, C. Gauthier, Polym. Adv. Technol. **1995**, 6, 351
[2] C.J. Lintner, J. Prak. Chem. **1886**, 34, 378.
[3] J.P. Robin, C. Mercier, R. Charbonnière, A. Guilbot, Cereal Chem. **1974**, 51, 389.
[4] A. Dufresne, J.Y Cavaillé., W. Helbert, Macromolecules **1996**, 29, 7624.
[5] J.L. Putaux, S. Molina-Boisseau, T. Momaur, A. Dufresne, Biomacromolecules **2003**, 4(5), 1198.
[6] O.A. Battista, In Microcrystal Polymer Science, Mc Graw-Hill Book Company, New York **1975**, 138.
[7] H.F. Zobel, Starch/Stärke **1988**, 40(1), 1.
[8] A.H. Muhr, J.M.V. Blanshard, D.R. Bates, Carbohydr. Polym. **1984**, 4, 399.
[9] D.M. Hall, J.G. Sayre, Textile Research **1970**, 40, 147.
[10] P. Kassenbeck, Starch/Stärke **1978**, 30(2), 40.
[11] G.T. Ooestergetel, E.F.G. van Brugen, Starch/Stärke **1989**, 41(9), 331.
[12] P.D. Gallant, A. Guilbot, Stärke **1969**, 21, 156.
[13] C. Sterling, Starch/Stärke **1974**, 26, 105
[14] H.C.H. Wu, A. Sarko, Carbohydr. Res. **1978**, 61, 27
[15] A. Buléon, B. Pontoire, C. Riekel, H. Chanzy, W. Helbert, R. Vuong, Macromolecules **1997**, 30, 3952
[16] D. French, in : Starch, Chemistry and Technology, R.L. Whistler, J.N ; Bemiller, E.F. Parschall editors, New York : Academic Press **1984**, 183.
[17] M. Yamaguchi, K. Kainuma, D.J. French, J. Ultrastruct. Res. **1979**, 69, 249.
[18] G. Hollinger, R.H. Marchessault, Biopolymers **1975**, 14, 265.
[19] D. Gallant, B. Bouchet, A. Buléon, S. Perez, Eur. J. Clin. Nutr. **1992**, 46, S3.
[20] P.J. Jenkins, R.E Cameron, A.M. Donald, Starch/Stärke **1993**, 45(12), 417.
[21] P.J. Jenkins, A.M. Donald, Int. J. Biol. Macromol. **1995**, 17(6), 315.

Structure Analysis of PVC Nanocomposites

Lucie Kovarova,[1] *Alena Kalendova,*[1] *Jean-Francois Gerard,*[2] *Jiri Malac,*[1] *Josef Simonik,*[1] *Zdenek Weiss*[3]

[1] The Department of Polymer Engineering, Faculty of Technology, Tomas Bata University in Zlin, TGM 275, 762 72 Zlin, Czech Republic
Fax: (+420) 57 603 1328; E-mail: lkovarova@ft.utb.cz
[2] Laboratoire des Materiaux Macromoleculaires UMR CNRS 5627/IMP, INSA, Lyon, France
[3] Central Analytical Laboratory, Technical University Ostrava, 708 33 Ostrava-Poruba, Czech Republic

Summary: In this paper poly(vinyl chloride)/clay nanocomposites were prepared by melt intercalation using a single screw extruder. Problems with thermal stability of these nanocomposites during compounding were largely eliminated by pre-treatment of the organoclay with plasticizer (dioctyl phthalate), which created a barrier between polymer and quaternary amine. These nanocomposite materials were analyzed with respect to their morphology. The intercalation, exfoliation, nano-phase dispersion and orientation were investigated using Transmission Electron Microscopy (TEM), Scanning Electron Microscopy (SEM), Atomic Force Microscopy (AFM) and X-ray diffraction (XRD). Moreover, different types of sample preparation for these techniques were tested as well. It was found that partially intercalated and disordered structure arose in poly (vinyl chloride) composites containing sodium type of montmorillonite, while a fine dispersion of partial to nearly full exfoliation of individual montmorillonite layers in poly (vinyl chloride) matrix was observed when this clay was organically modified. Finally, the influence of different mixing time (in extruder) on nano-phase morphology was surveyed.

Keywords: melt intercalation; montmorillonite; morphology; nanocomposites; poly(vinyl chloride)

Introduction

Clays have long been used as fillers in polymer systems because of their low cost and improved mechanical properties of the resulting polymer composite. However, clay particles could only be dispersed on the micro-scale. In the early 1990`s, Toyota researchers discovered that the treatment of montmorillonite clay with amino acids allowed the dispersion of individual 1 nanometre thick silicate layers of the clay on a molecular scale in polyamide-6, which was one of the latest evolutionary steps in polymer technology [1]. Therefore, many recent studies have focused on polymer-layered silicate

 DOI: 10.1002/masy.200550311

nanocomposites, which can show very large improvements in mechanical, thermal and barrier properties, as well as in dimensional stability and remarkable increase in heat distortion temperature without a significant loss of transparency, toughness or impact strength [2, 3, 4]. The enhancement of these properties mainly depends on the interaction between the polymer matrix and the layered silicates. According to this interaction quality, the different types of nanocomposites ranging from intercalated to exfoliated states may be obtained. Thus, the radical improvement is ascribed to the nano-scale morphology, which means homogeneous dispersion and exfoliation of individual silicate layers in the continuous polymer matrix [5].

A great number of published papers deal with intercalation of monomers or polymers into layered silicate clay as one of the promising strategies for nanocomposites arrangement. In most cases, this synthesis involves either intercalation of a suitable monomer and then exfoliation of the layered silicate clay into individual nano-scale elements by subsequent polymerization or polymer intercalation from a solution. However, for majority of commodity polymers, both approaches are limited since neither a suitable monomer nor a compatible solvent is always available [5, 6]. Therefore, polymer-melt direct intercalation method can be a new approach to fabrication of polymer-layered silicate nanocomposites by using a conventional polymer extrusion process. In this particular case, the formation of nanocomposites depends upon the thermodynamic interaction between the polymer chains and the host silicate surfaces, and also upon the transport of the polymer chains from the bulk melt into the layered silicate clay. It was found that the result of polymer intercalation is determined by the interplay of entropic and enthalpic factors [5]. In fact, although the motion reduction of polymer chains inside the interlayer of silicate clay mineral results in a decrease in the overall entropy of macromolecular chains, this entropic decrement may be compensated by an increase in conformational entropic freedom of the tethered alkyl surfactant chains, which separate the silicate layers of clay. Since small increases in the basal spacing of layered silicate do not strongly influence the total entropic change, intercalation will rather be driven by the changes in the total enthalpy. Therefore, the driving force of this mechanism is the enthalpic contribution of interactions between polymer and organically modified clay [6, 7].

The literature contains numerous studies about melt intercalation method for nanocomposites preparation where polyamide, polystyrene, polypropylene, polyethylene and epoxy resin are mainly used as polymer matrix [8]. Only recently, polyvinylchloride nanocomposites have been prepared by polymer melt intercalation. However, few publications about PVC containing montmorillonite as a layered silicate clay have been advertised so far, despite the fact that polyvinylchloride is an important commercial polymer owing to its wide applications and low cost [9, 10, 11, 12, 13]. Wang et al. [11] primarily studied the influence of plasticizer (in this case dioctyl phthalate) and clay amount on thermal and mechanical properties of the nanocomposites. Their work contained several kinds of intercalated PVC nanocomposites. On the other hand, they were not interested in morphology, because XRD was the only method for observing the structure of their PVC nanocomposites. Wang et al. [12] mainly investigated the effect of silicate clay modification and its content on the relaxation behaviour, mechanical properties and morphology. In this publication XRD and TEM were two methods used for PVC nanocomposite structure observation.

None of the publications about PVC nanocomposites include techniques such as Scanning Electron Microscopy and Atomic Force Microscopy. However, these techniques are very important to characterize the resulting nanocomposites morphology and a lot of researchers use them together with TEM and XRD to confirm nanocomposites structure. Therefore, the aim of the present paper is to give information about the morphology of PVC nanocomposites using four basic methods: TEM, SEM, AFM and XRD. Two types of PVC/MMT nanocomposites were prepared by polymer melt intercalation with sodium type of montmorillonite and organically modified montmorillonite, in which dioctyl phthalate was applied as a plasticizer. In addition, the influence of two different retention times (single and double pass compounding) in extruder on PVC nanocomposites morphology is discussed.

Experimental

Materials

Two types of smectic clay, namely Cloisite Na+ and Cloisite 30B, from Southern Clay Products, USA were used in this study. The first one is sodium montmorillonite without any surface treatment, the second one is a commercial product modified by methyl tallow bis-2-hydroxyethyl ammonium chloride at a loading of 90 milliequivalent per 100 grams of neat clay to form the lamellar clay/tallow core layered structure. As a polymer matrix the suspension type of polyvinylchloride (PVC) Neralit 652 (K-value 65) supplied by Spolana Neratovice, Czech Republic, was utilized. Dioctyl phthalate (DOP) was used as a

PVC plasticizer. All samples were stabilized with 2 wt. % of Lankromark LZB 968, which is quite often industrially used in the field of flexible PVC flooring. The main characteristics of the PVC composition used in this study are listed in Table 1.

Table 1. Main characteristics of the PVC compound.

Component	*Commercial name*	*Concentration*
Polyvinylchloride	Neralit 652 (K-value 65)	74 wt. %
Plasticizer	Dioctyl phthalate (DOP)	24 wt. %
Stabilizer	Lankromark LZB 968	2 wt. %

Sample preparation

First co-intercalation of Cloisite Na+ and Cloisite 30B by dioctyl phthalate was carried out at the temperature of 80 °C in a laboratory mixer for 30 minutes. In the next step PVC blend containing plasticizer, stabilizer and also modified layered silicate (5 wt. %) was mechanically mixed and thereafter compounded in Buss KO-kneader MKS 30 single screw extruder. The polymer melt intercalation method was used as a promising new approach to fabricate PVC nanocomposites using polymer processing technique. The amount of nanofiller was calculated to be 5 wt. % of pure sodium montmorillonite (inorganic material) in the whole mixture. Therefore, the amount of Cloisite 30B was recalculated with respect to the amount of organic modificator. The operating temperature was from 130 to 160 °C and the screw speed 35 rpm. The conditions of mixing were same for all samples. The effect of different retention time of compounding on the morphology was investigated as well. Therefore, all compounds (PVC + nanofiller + plasticizer+ stabilizer) were compounded twice (double-pass compounding) in single screw extruder.

Characterization and Testing

The nano-phase orientation, dispersion, intercalation and exfoliation of both clays in PVC matrix were observed through four main microscopic techniques.

The degree of exfoliation and intercalation was studied using reflection mode on URD-6 Diffractometer with $CuK\alpha_1$ radiation (λ = 1.54 Å). This method was used not only for PVC nanocomposites, but also for the powder form of both nanofillers.

For transmission electron microscopy (TEM, Phillips CM120) the specimens were prepared by cutting on a cryoultramicrotome equipped with a diamond knife. The cutting temperature was not same for every sample, but it was set during cutting with respect to the sample behavior, and ranged from -30°C to -5°C.

For scanning electron microscopy (SEM, JEOL JSM-35) the samples were prepared through two different techniques. In the fist one the samples were placed in liquid nitrogen and freeze-fractured, the second samples were cut at room temperature on an ultramicrotome and then plasma-etched using oxygen plasma. Plasma-etching device with the power of 10W at the frequency of 13.56 MHz was used.

The sample preparation for Atomic Force Microscopy (EXPLORER™, THERMOMICROSCOPES) was the same as the second method of SEM sample preparation using oxygen plasma etching. Plasma treatment is selective, because mineral particles of montmorillonite are inactive to the etching process, while PVC is very easily burned off.

Results and Discussion

Figure 1 shows the maxima of two nanofillers, Cloisite Na^+ (nature sodium type of montmorillonite) and Cloisite 30B (intercalated sodium type of montmorillonite). As can be seen, the former exhibits its characteristic diffraction peak at approximately 7.8°, which corresponds to basal spacing of 1.2 nm. The latter indicates its peak at 4.83°, which is equivalent to the basal spacing of 1.85 nm. This higher basal spacing compared with the value of Cloisite Na^+ is caused by intercalation of organic compound into interlayer spaces of montmorillonite.

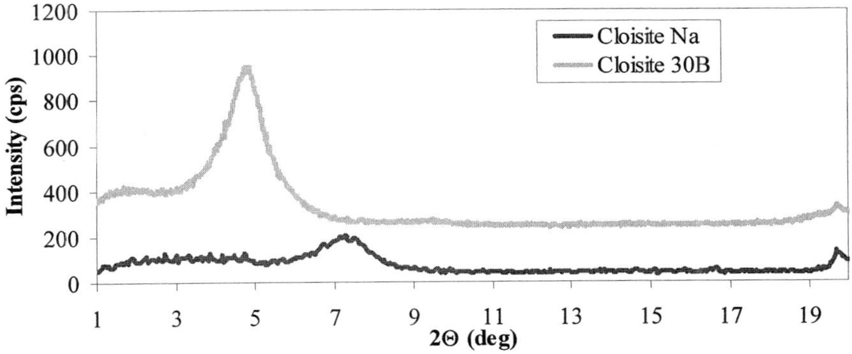

Figure 1. XRD patterns of Cloisite Na^+ and Cloisite 30B.

Our preliminary technological testing demonstrated that the polymer stability is greatly influenced by the used type of nanofiller, especially the type of organic compound utilized as an intercalating agent. Our idea was to cover the quaternary amine groups with plasticizer and therefore prevent a contact between these amine groups and active chlorine atoms. In the case of Cloisite 30B the quaternary amine groups can be the source of degradation process. Therefore, in the first step we co-intercalated the plasticizer (dioctyl phthalate) into organically modified Cloisite 30B. Gradually, these silicate nanofillers (Cloisite Na^+ and plasticizer-modified Cloisite 30B) were dispersed in polyvinylchloride matrix.

As can be seen from Figure 2, which shows XRD patterns of PVC composites containing Cloisite Na^+, there is a shift of peaks maximum towards the lower 2Θ value in the case of longer retention time. It means that the larger basal spacing (silicate layer plus interlayer space) corresponding to the lower value of the peaks maximum can be induced by the longer time of mixing in Buss KO-kneader. It can support the transport of polymer chains from the bulk melt into the interlayer space of layered silicates. On the other hand, PVC composites with Cloisite Na^+ did not show remarkable changes in the peaks maxima compared with the data of pure Cloisite Na^+ powder in Figure 1. It signifies that a structure containing agglomerates of the nanoparticles was obtained.

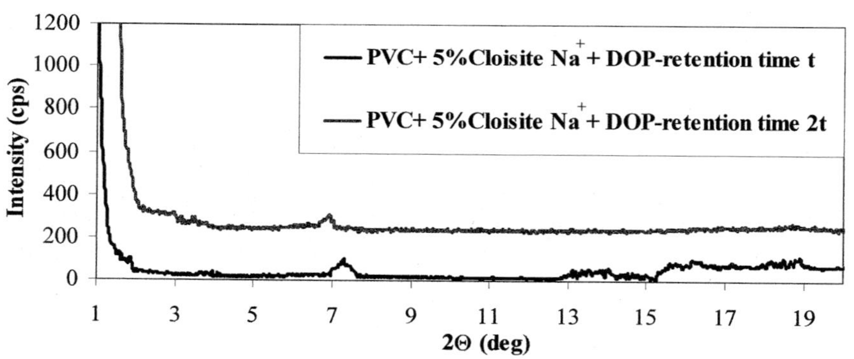

Figure 2. XRD scans of PVC/Na+ MMT composites with different time of mixing in Buss KO-kneader.

In order to get a more precise interpretation, the morphology of PVC/clay nanocomposites was also examined by TEM technique. Figures 3 and 4 illustrate the structure of these samples. The dispersion of clay particles (Na^+ MMT) in PVC matrix has resulted in the formation of clay tactoides with the layers aggregated in an unintercalated face-face form. The longer retention time has caused better dispersion of clay tactoides in PVC matrix. However, it does not significantly effect the intercalation and exfoliation of the clay particles.

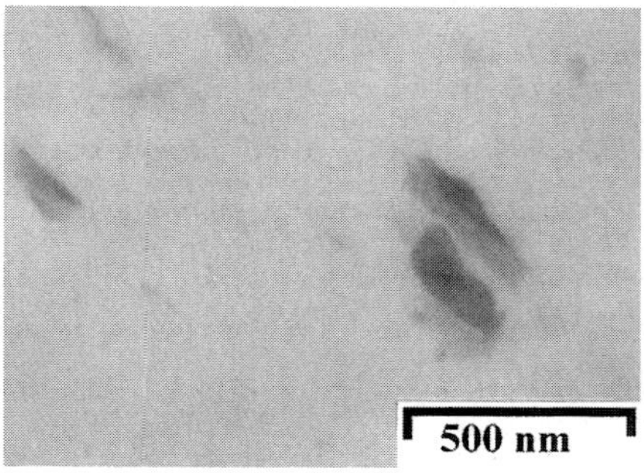

Figure 3. TEM micrograph of intercalated PVC/Na^{+} MMT nanocomposite (retention time t).

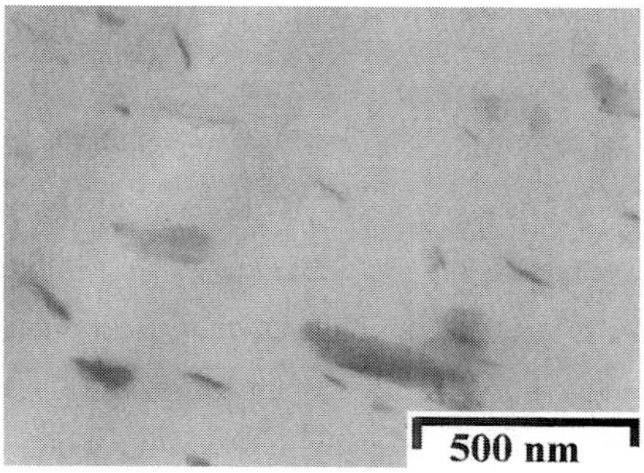

Figure 4. TEM micrograph of intercalated PVC/Na+ MMT nanocomposite (retention time 2t).

As revealed the Figure 5, the peaks maxima of PVC + Cloisite 30B + DOP are almost the same for two samples containing Cloisite 30B. They show two characteristic peaks at 2Θ values of 2.1 and 4.64°, corresponding to basal spacing of 4.3 and 1.9 nm, respectively. As can be seen from a comparison of the peaks maxima for Cloisite 30B in Figure 1 (powder form of Cloisite 30B) and 5 (PVC composites containing Cloisite 30B), there are new peaks corresponding to higher basal spacing in the case of Cloisite 30B dispersed in PVC matrix. The finite expansion of gallery height (basal spacing) is probably associated with the intercalation of organic compound and also polymer chains into clay interlayer since the larger distance of montmorillonite layers obviously demonstrates the advantage of the intercalation or co-intercalation process, the better dispersion of clay in polymer matrix or the higher level of exfoliation.

This interpretation can be evidenced by TEM technique in Figures 6 and 7. Individual clay platelets are clearly visible. These two pictures demonstrate that many silicate layers are exfoliated into nanometer layers. It is very probable that the longer time of mixing in Buss KO-kneader causes regular dispersion and orientation of the nanoparticles in the PVC matrix via melt-intercalation process.

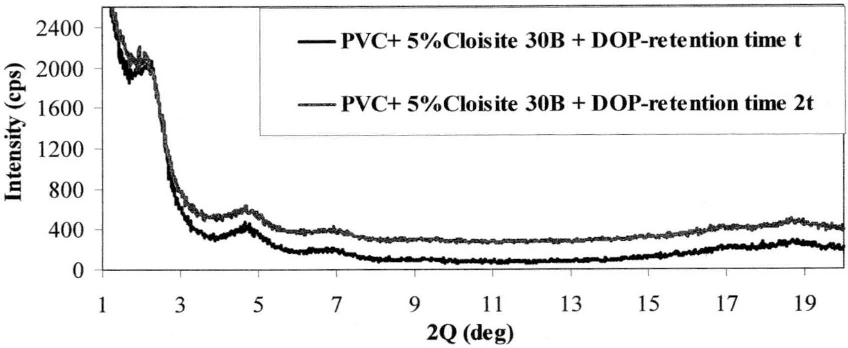

Figure 5. XRD scans of PVC/Cloisite 30B nanocomposites with different time of mixing in Buss KO-kneader.

Figure 6. TEM micrograph of PVC/Cloisite 30B nanocomposite (retention time t).

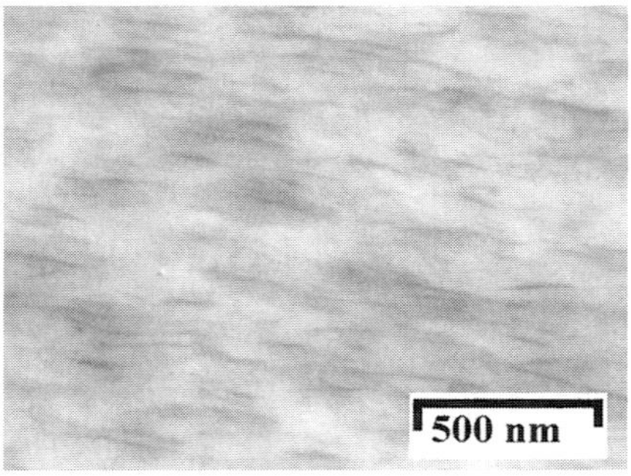

Figure 7. TEM micrograph of PVC/Cloisite 30B nanocomposite (retention time 2t).

Figures 8 and 9 show PVC/Cloisite 30B nanocomposite (retention time t) obtained by scanning electron microscopy for comparison with transmission electron microscopy. They differ in the method of sample preparation. In the first case the common technique was used, where the sample was placed in liquid nitrogen and then freeze fractured. The second sample, however, was cut using an ultramicrotome at room temperature and then plasma-etched using oxygen. The sample prepared only by cutting (on ultramicrotome) was not suitable for SEM, because individual clay particles were not distinguished from polymer (PVC) matrix. The results given in both figures support the idea that very good dispersion and intercalation of clay particles in PVC matrix was received. The silicate layers are better visible in Figure 9, where polymer matrix was burned due to the oxygen plasma. On the other hand, there might have been a lot of other structures or artefacts, in other words significant morphological changes, caused by burning away the polymer.

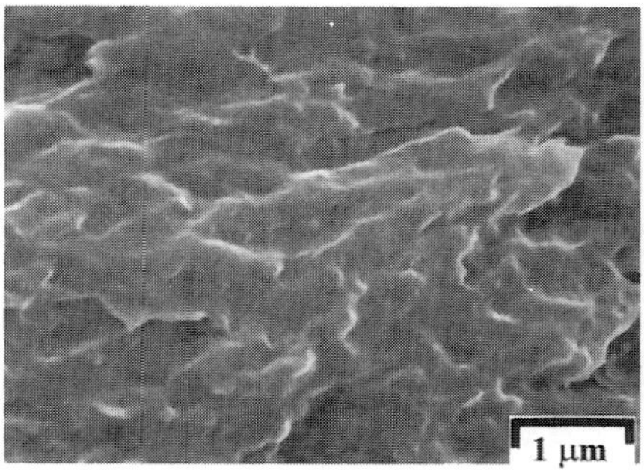

Figure 8. SEM micrograph of PVC/Cloisite 30B nanocomposite – retention time t (the samples were placed in liquid nitrogen and freeze-fractured).

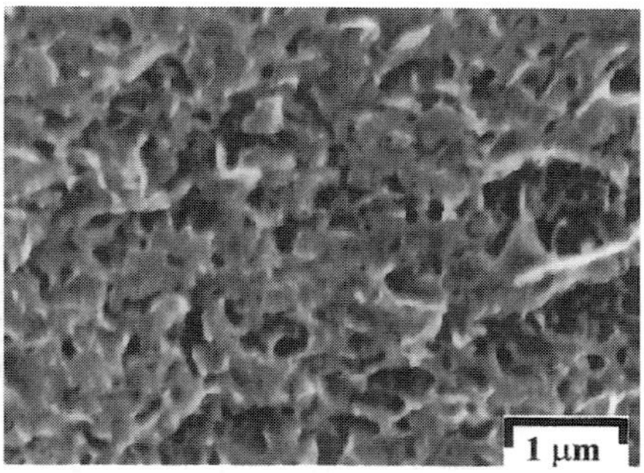

Figure 9. SEM micrograph of PVC/Cloisite 30B nanocomposite – retention time t (the sample was cut on an ultramicrotome and then plasma etched).

Scanning Electron Microscopy in Transmission Mode can provide very similar micrograph compared with TEM technique; this is shown in Figure 10 of PVC/Cloisite 30B nanocomposite (retention time 2t) prepared by cutting on an ultramicrotome at room temperature. There is a slight possibility that in the case of cutting at room temperature larger-scale deformation of thin clay particles originated compared with cutting at low temperature (freeze cutting). However, this room temperature cutting is easier and cheaper. Figure 11 shows AFM micrograph of PVC/Cloisite 30B nanocomposite (retention time 2t) obtained in the non-contact mode. The lack of detectability between individual clay particles and polymer matrix in the case of using only cutting at room temperature resulted in utilizing oxygen plasma etching. So, sample preparation was the same as the second sample for SEM.

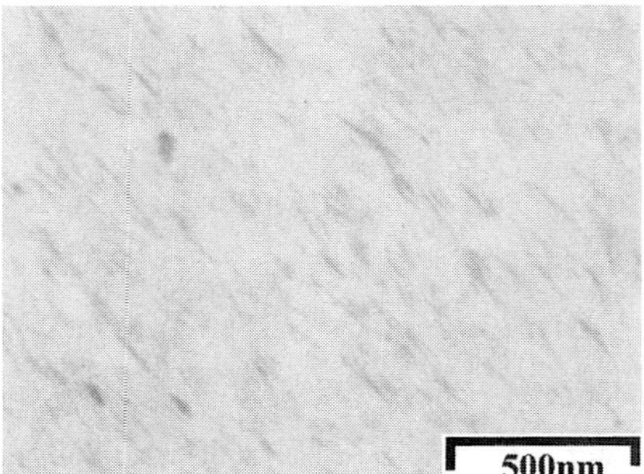

Figure 10. STEM micrograph of PVC/Cloisite 30B nanocomposite – retention time 2t (the sample was cut on an ultramicrotome).

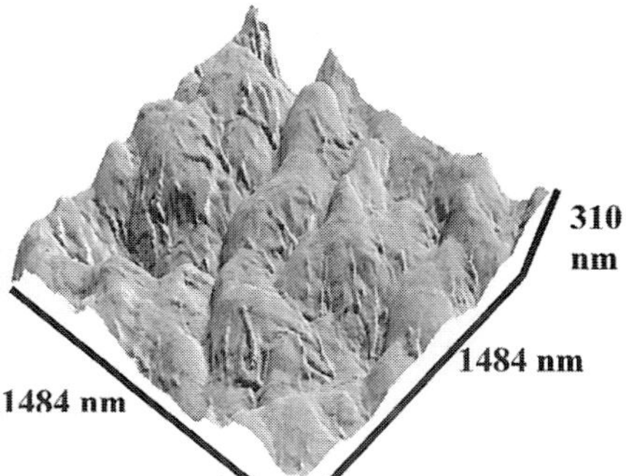

Figure 11. AFM micrograph of PVC/Cloisite 30B nanocomposite (the sample was cut on an ultramicrotome and then plasma-etched).

Conclusions

The objective of this study was to compare different types of methods to explore morphology of PVC nanocomposites. TEM, SEM, AFM and X-ray diffraction were the basic methods used. Partially intercalated and disordered PVC nanocomposites were obtained by melt compounding of PVC with sodium type of montmorillonite (Cloisite Na^{+}) in the presence of DOP. However, the same process resulted in intercalated and partially exfoliated structures of PVC nanocomposites with organically modified montmorillonite (Cloisite 30B). It was found that the longer retention time (time of sample mixing) in Buss KO-kneader can induce the better dispersion and orientation of silicate nanoparticles in PVC matrix. On the other hand, it is highly improbable that the longer time of compounding noticeably affects the exfoliation process without previous intercalation of montmorillonite. It should be noted that this study has examined only morphology of PVC nanocomposites. Hence, further detailed studies on the determination of mechanical, thermal and rheological properties of PVC nanocomposites were done and will be published in our next article. Unfortunately, we are unable to say if the sample preparation in each microscopic method can be suitable for other polymers to obtain good results (clear micrograph).

Acknowledgments

We would like to thank all people, especially J. Duchet, J. Galy and A. Rivoire from the Laboratory of Macromolecular Materials, INSA, Lyon, France, A. Lengalova from Faculty of Management and Economics, Tomas Bata University in Zlin, Czech Republic and M. Kristkova from the Central Analytical Laboratory, Technical University Ostrava, Czech Republic for their help.

[1] Y. Fukushima, A. Okada, M. Kawasumi, T. Kurauchi, O. Kamigaito, *Clay Mineralogy* **1988**, *23*, 27.
[2] L. Berlund, *Fillers and Additives for Plastics* **2000**, *4*, 3.1.
[3] E.P. Gianellis, *Adv. Mater* **1996**, *8* (1), 29.
[4]M. Kato, A.Usuki, in "*Polymer-Clay Nanocomposites*", T. J. Pinnavaia, G. W. Beall, Eds:, J. Wiley & Sons, England, 2000, p. 5.
[5] M. Alexandre, P. Dubios, *Mater. Sci. Eng.* **2000**, *28*, 1.
[6] R. A. Vaia, E. P: Giannelis, *Macromolecules,* **1997**, *30*, 7990.
[7] L. Liu, Z. Qi, X. Zhu, *J. Appl. Polym. Sci.* **1999**, *71*, 1133.
[8] Y. H. Zhang, K. C. Gong, *Mat. Res. Soc.* Symp. Proc. **1998**, 520, 191.
[9] J. Trlica, A. Kalendova, Z. Malac, J. Simonik, L. Pospisil, *ANTEC 2001*, Conference Proceedings, Dallas, USA, May 6-10, **2001**, 2162.
[10] L. Kovarova, A. Kalendova, J. Malac, J. Simonik, *PPS -19*, Conference Proceedings, Melbourne, Australia, July 7-10, **2003**.
[11] D. Y. Wang, D. Parlow, C. A. Wilkie, *Polym. Prepr.* **2001**, *42* (2), 842.
[12] CH. Wan, X. Qiao, Y. Zhang, Y. Zhang, *Polym. Test.* **2003**, *22*, 453.
[13] J. Du, D. Wang, Ch. A. Wilkie, J. Wang, *Polym. Degrad. Stabil.* **2003**, *79*, 319.

Polymer Layered Silicate/Carbon Nanotube Nanocomposites: Morphological and Rheological Properties

S. Peeterbroeck,[1] *M. Alexandre,*[1] *J. B. Nagy,*[2] *N. Moreau,*[2] *A. Destrée,*[2] *F. Monteverde,*[3] *A. Rulmont,*[4] *R. Jérôme,*[5] *Ph. Dubois**[1]

[1] Service des Matériaux Polymères et Composites, Université de Mons-Hainaut, Place du Parc 20, B-7000 MONS, Belgium
[2] Laboratoire de Résonance Magnétique Nucléaire, Facultés Universitaires Notre-Dame de la Paix, Rue de Bruxelles 61, B-5000 NAMUR, Belgium
[3] Unité de Microscopie Electronique, Materia Nova Center, Parc Initialis, Avenue N. Copernic 1, B-7000 MONS, Belgium
[4] Laboratoire de Chimie Inorganique Structurale, Université de Liège, Bâtiment B6a, B- 4000 LIEGE, Belgium
[5] Centre d'Etude et de Recherche sur les Macromolécules, Université de Liège, Bâtiment B6a, B- 4000 LIEGE, Belgium

Summary: Morphological and rheological properties of new ternary nanocomposites based on ethylene vinyl acetate copolymers (EVA), commercial organo-modified clays (organoclays) and purified multi-walled carbon nanotubes (MWNTs), prepared via direct melt blending, have been evaluated. For sake of comparison, the corresponding binary compositions, i.e., EVA filled with either organoclays or MWNTs, have been investigated as well. While extensive exfoliation can be observed for binary EVA/clay nanocomposites, the addition of MWNTs appears to limit clay exfoliation. Rheological properties show that both clay and MWNTs increase the elastic modulus of the nanocomposites, reflecting the high degree of nanoparticle interconnectivity that can be found in these materials.

Keywords: carbon nanotubes; clay; EVA; nanocomposites; rheology

Introduction

Homogeneously dispersing rigid particles in polymer matrices is a very common and widely used method that allows for readily increasing the stiffness of the so-obtained composite materials. Depending on the intrinsic nature of the added filler, other properties can also be enhanced such as fire resistance, electrical and thermal properties. However, such improvements of composite materials performances usually require high filling levels [1], detrimental to the ultimate mechanical properties of the resulting materials.

 DOI: 10.1002/masy.200550312

Polymer nanocomposites represent a new class of composite materials, i.e., particle-filled polymers for which at least one dimension of the dispersed particles is in the nanometer range. This variety of fillers provides the related composite materials with significantly improved properties most often at filler content as tiny as 3 to 5 wt%. Interestingly, remarkable thermo-mechanical performances have been recorded using organo-modified layered silicates (organoclays) as nanofiller precursors [2-4]. Actually, upon incorporation of such organoclays in a polymer matrix, intercalation of the polymer chain in-between the clay platelets or complete exfoliation of the silicate platelets can be observed. Exfoliated nanocomposites usually exhibit significantly higher improvement in properties such as mechanical, thermal and flame retardant properties.

First observed by Ijima in 1991 [5], carbon nanotubes is another family of potential nanofillers for polymer matrices. Carbon nanotubes are mainly subdivided in two families: single-walled nanotubes (SWNTs) or multi–walled nanotubes (MWNTs) where several nanotubes of decreasing diameter are interlocked. Carbon nanotubes find applications in various fields such as in field emission devices, electrically and thermally conductive materials, hydrogen storage devices and molecular sieves [6-8]. Even though carbon nanotubes have already been blended within various polymers [9-11], homogeneous dispersion of the nanotubes in the polymer matrix remains one major challenge, since bundles of aggregated SWNTs or MWNTs most often persist and therefore limit the performances of the recovered composite materials. Interestingly, a synergistic effect arising from the combination of carbon nanotubes and organoclays has been recently evidenced [12,13]. Actually, when selected organoclays and MWNTs were blended simultaneously within a soft ethylene vinyl acetate copolymer (EVA) in the molten state, both thermal and flame retardant properties of the so-recovered ternary compositions proved to be much enhanced. In order to shed some light on the origins of such a synergistic effect, it was of prime interest to characterize these nanocomposites and to understand the parameters governing such an original behavior.

In this context, the present work aims at identifying the (nano)morphology of EVA-based compositions filled with selected organoclays and/or purified MWNTs and obtained by direct melt blending. In addition to morphological characterization, the different binary and ternary nanocomposites have been also compared from the point of view of their rheology.

Experimental Part

Materials

A commercial ethylene vinyl acetate copolymer (EVA), Escorene UL00328 (Exxon), containing 27wt% of vinyl acetate, was chosen as the matrix. A commercial organoclay, Cloisite®30B, modified by 20.3 wt% of bis(2-hydroxylethyl)methyl tallowalkyl ammonium cation (Southern Clay Products, USA) was used. This montmorillonite has a specific surface area of ca. 750 m^2/g, with platelets of a thickness around 1 nm. The MWNTs used in this work were produced by catalytic decomposition of acetylene on transition metal particles (Co, Fe) supported on Al_2O_3. The catalyst contains 2.5 wt% cobalt and 2.5wt% iron supported on alumina [13]. The synthesis was carried out in a fixed-bed flow reactor at 700°C for reaction time of 60 min. Purified MWNTs were obtained after dissolution of the support in boiling concentrated NaOH water solution and dissolution of the catalysts in concentrated hydrochloric acid water solution. The MWNTs are characterized by an average inner diameter of 5 nm and an average outer diameter of 15 nm, corresponding to ca. 14 - 15 concentric layers. The average length is ca. 50 μm.

Preparation

EVA/organoclay nanocomposites were compounded on an AGILA two-roll mill for 12 min at 140°C with a friction coefficient equal to 1.35 and a rotation speed of 15 m/s for the slowest roll. EVA/MWNT and EVA/organoclay/MWNT nanocomposites were prepared in a Brabender internal mixer at 140°C, for 12 min with a screw speed of 45 rpm.

Characterization

XRD was measured using a Siemens D5000 diffractometer with the Cu Kα radiation (λ = 0.15406 nm) from 1.5 to 30° by step of 0.04°. Transmission electron micrographs were obtained with a Philips CM200 apparatus using an accelerator voltage of 100 kV. The samples were 70-80 nm-thick and prepared with a LEICA Ultracut UCT ultracryomicrotome cutting at –130°C. Rheological measurements in the melt were obtained in oscillatory mode using disk and plate geometry in a Rheometrics Dynamic Analyser. The measurements were performed at 140°C on disks (diameter = 25 mm and thickness = 3 mm), with a frequency sweep from 15.9 Hz to 0.01 Hz, and 1% strain.

Results and Discussion

Organo-modified montmorillonites such as Cloisite®30B are known to form nanocomposites with improved fire retardant properties and increased stiffness when dispersed in EVA [14-17]. In a previous study, thermal degradation and tensile properties of binary and ternary blends of EVA with Cloisite®30B and/or MWNTs have been investigated [12]. The properties of the resulting new binary and ternary nanocomposites were enhanced by the presence of both nanofillers. A synergistic effect was observed when the organoclay and MWNTs were added simultaneously: the thermal and flame retardant properties of the related EVA matrices were enhanced.

Morphological characterization

It was of interest to explore the morphology of these new materials. X-ray diffraction (XRD) and transmission electron microscopy (TEM) were used to identify the nanostructure of these EVA-based composites.

Figure 1 compares the XRD patterns for Cloisite®30B, EVA matrix and EVA-organoclay nanocomposites filled with 1, 3, or 5wt% silicate platelets. The pattern of the pure EVA presents a large peak centered at $2\theta = 20°$, characteristic of the crystallinity of the matrix. Although the Cloisite®30B shows a peak centered at $2\theta = 4.8°$, characteristic for an interlayer spacing of 18.5 Å, the patterns of the composites based on this organoclay do not show any diffraction peak in the 1.5 – 10° angle region for a filler content below 5wt%. This indicates the disruption/loss of the repetitive structure and thus possibly the exfoliation of the silicate layers. At 5wt% a poorly defined shoulder is observed in the low 2θ region indicating that some intercalation might remain. Confirming XRD data, the sample containing 3wt% of clay displays a fully exfoliated morphology as observed by TEM (Figure 2). Homogeneously dispersed single clay layers are found in abundance throughout the polymer such that an XRD signal at low 2θ angle would not be expected.

The XRD patterns for EVA-purified MWNTs composites have also been recorded. Whereas such carbon nanotubes should be characterized by diffraction peaks at $2\theta = 25.8°$ and 42.8° [11], no diffraction peak was observed in that region (XRD pattern not shown here). TEM micrographs

(Figure 3) show a non homogeneous dispersion of the carbon nanotubes in the EVA matrix. The majority of them are interconnected, exhibiting a high degree of particle-particle entanglements.

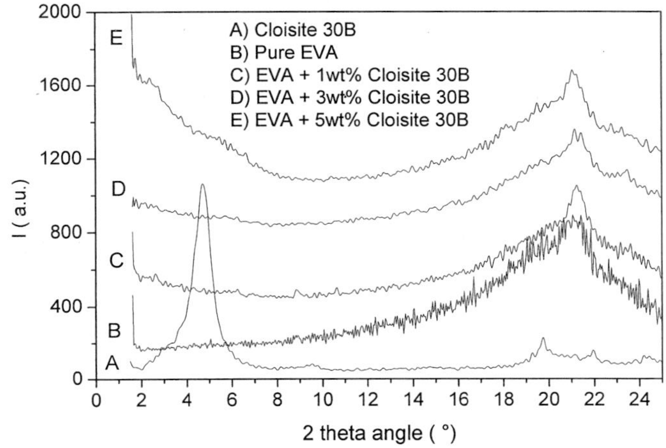

Figure 1. XRD patterns for EVA, Cloisite®30B and binary compositions containing EVA and various amounts of Cloisite®30B.

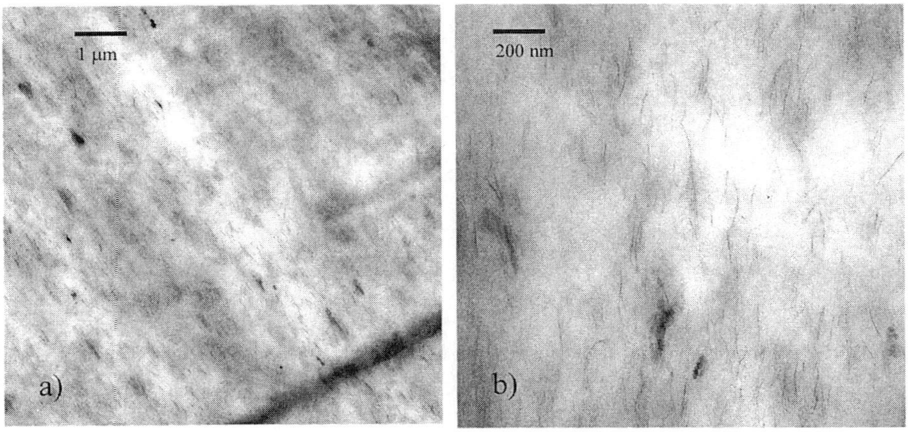

Figure 2. TEM micrographs of a binary system containing EVA and 3wt% Cloisite®30B a) low magnification , b) high magnification.

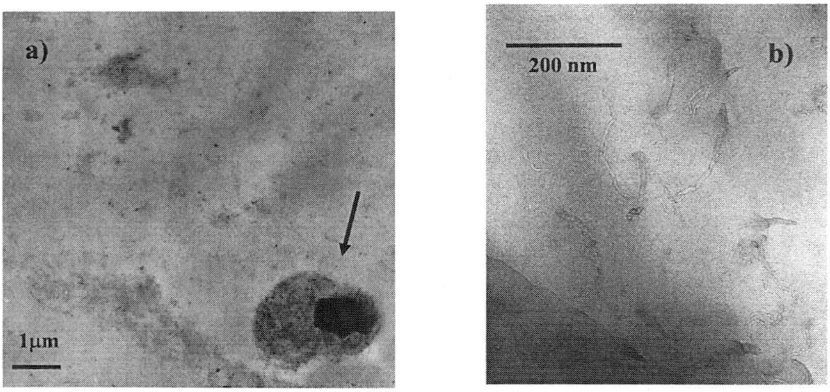

Figure 3. TEM micrographs of a binary system containing EVA and 3wt% MWNTs: a) low magnification with the arrow showing a large aggregate (top view of a MWNTs bundle) b) high magnification showing interconnected MWNTs.

Figure 4 presents the XRD patterns of two ternary nanocomposites both containing 1wt% purified MWNTs and either 2 or 3wt% Cloisite®30B. A diffraction peak in the region of $2\theta = 6°$ is detected this time for both ternary materials, however more intense in the presence of 3wt% organoclay. A semi-exfoliated semi-intercalated structure is more likely formed despite of the limited quantity of filler (< 5wt%). Carbon nanotubes might be responsible for that observation as they might perturb the good dispersion/delamination of Cloisite®30B in the EVA matrix. Indeed, it is worth reminding that the XRD pattern of the corresponding binary system (again 3wt% Cloisite®30B), thus presented in Figure 1, did not show any diffraction peak at low 2θ angle. When MWNTs are added with the organoclay, the homogeneity of the platelets dispersion and the extensive exfoliated structure are lost, even at low loading level.

TEM micrographs (Figure 5) of the ternary composite containing 3wt% Cloisite®30B and 1wt% MWNTs confirm XRD observations. Indeed the clay is again exfoliated but not quantitatively. On the other hand, it appears that even tough the carbon nanotubes are not fully and homogeneously dispersed in EVA, one can distinguish some individual MWNTs (segments) throughout the sample, again showing rather high degree of interconnectivity. It is worth recalling that the investigated carbon nanotubes are characterized by an average length of 50μm.

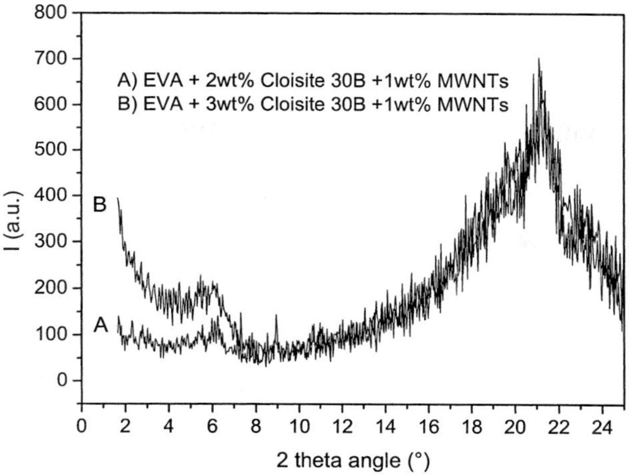

Figure 4. XRD patterns of ternary systems based on 1wt% MWNTs and 2 or 3wt% Cloisite®30B.

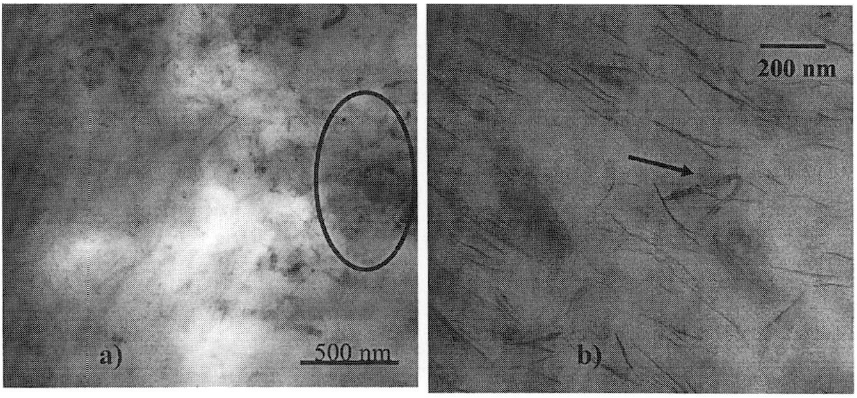

Figure 5. TEM micrographs of a ternary composite (3wt% Cloisite®30B+1wt% MWNTs). a) low magnification showing MWNTs aggregates b) high magnification showing stacks of clay and an isolated carbon nanotubes (a segment of which being highlighted by the arrow).

Rheological characterization

Binary systems

Figures 6a and 6b shows the frequency dependence of the storage modulus G' for various EVA/organoclay and EVA/MWNT binary nanocomposites, respectively. As expected, G' of the organoclays-based nanocomposites increases with the clay loading on the whole range of swept frequencies (Figure 6a). At high frequencies, the qualitative behavior of G' is essentially the same whatever the silicate content (all data points turn into parallel lines). However at lower frequencies, G' increases monotonically with increasing clay content.

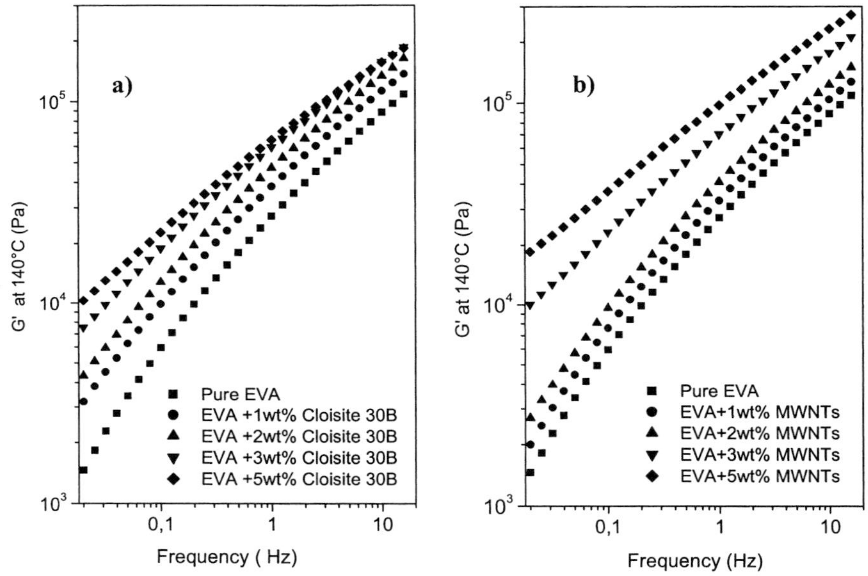

Figure 6. Frequency dependence of the storage modulus G' for binary nanocomposites at 140°C : a) EVA + Cloisite®30B, b) EVA + MWNTs.

Above 2wt%, the slope of the storage modulus curve gets less frequency dependent in the whole frequency range. This behavior is characteristic of a non-terminal frequency behavior [18]. This non-terminal behavior, which has also been observed for polystyrene nanocomposites [19] is attributed to the retardation of molecular relaxation processes produced by the confined geometric effect induced by the presence of highly dispersed nanoclay platelets. Thus a similar observation can be made for the EVA/organoclays nanocomposites. The finely dispersed silicate layers organize in a spatially-linked structure that impedes their free rotation, limiting therefore the relaxation of the material at low deformation frequencies.

For binary nanocomposites based on the purified MWNTs (Figure 6b), G' increases again with the carbon nanotube content. As the MWNTs content increases in the system and as a consequence of their length (ca. 50μm), nanotube-nanotube interactions should dominate more and more, eventually leading to percolation and the formation of an interconnected structure of nanotubes. At high concentrations of MWNTs (>2wt%), connectivity should get more pronounced, as seen in the enhanced elasticity. These particle-particle entanglements give to the material a better resistance against the applied deformation and a substantial increase in G' is that time observed even at higher frequencies (to be compared to EVA/clay nanocomposites in Figure 6a). The difference observed between the two binary nanocomposites above a filler content of 2wt%, is more likely due to the different material morphology and aspect ratio presented by the dispersed silicate platelets and entangled MWNTs.

Ternary systems

Figure 7 shows the frequency dependence of G' evolution recorded at 140°C for binary and ternary compositions with different loading levels. The rheological behavior of the ternary systems will be discussed in two separate sections: G' evolution for ternary composites will be compared first with binary systems containing the organoclay and secondly, with binary systems containing the purified MWNTs.

As far as the nanocomposites containing 3wt% Cloisite®30B are concerned, addition of nanotubes to organoclay leads to a clear modification in the G' evolution and a significant increase of the storage modulus G' in the low frequencies region (Figure 7a). This can be explained by the particle-particle entanglements present in the material after the addition of and triggered by the MWNTs. One could therefore assume that the association between such entanglements and the aforementioned spatially-linked structure gives to ternary nanocomposites a better resistance against low deformation. At higher frequencies, the G' values observed for the ternary composites containing 2wt% Cloisite®30B and 1wt% MWNTs are similar to the binary material only filled with 2wt% Cloisite®30B. This might result from the limited amount of clay in the ternary composite, likely too low for forming a spatially-linked structure at high frequencies. This is confirmed when looking at the frequency dependent G' evolution recorded for the ternary nanocomposite containing 3wt% Cloisite®30B and 1wt% MWNTs (Figure 7a). The slope of the curve is relatively constant and the values of G' are increased in the whole frequency range.

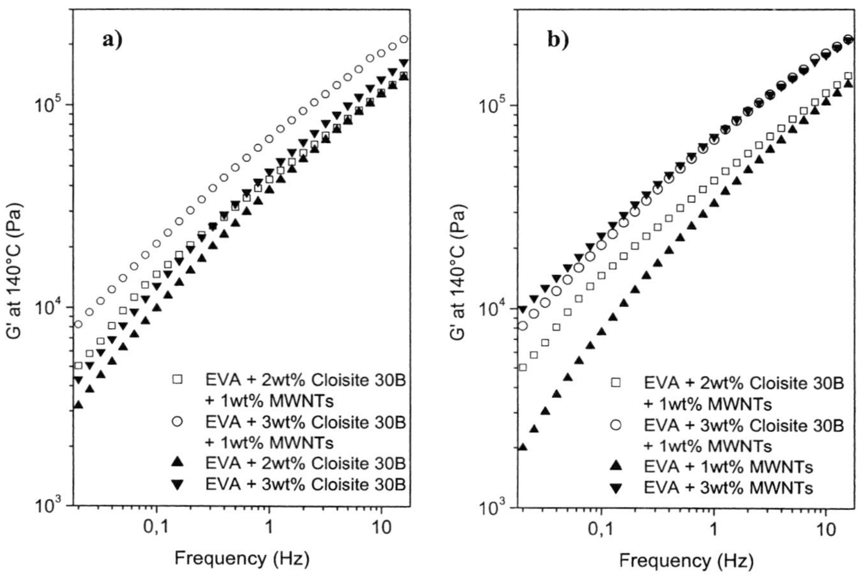

Figure 7. Frequency dependence of G' evolution at 140°C for binary and ternary systems a) comparison with EVA/organoclay binary compositions; b) comparison with EVA/MWNTs binary compositions.

Secondly, for the nanocomposites containing 3wt% of MWNTs, addition of clay platelets leads to a decrease of G' values in the whole frequency range as well as some modification in the slope of the evolution curve (Figure 7b). The conclusions are globally the same as above: a sufficient number of nanotube entanglements and a spatially-linked structure built by the homogeneously dispersed organoclays are required to give a good resistance to the material against high deformations. When the ternary composite contains 3wt% Cloisite® 30B and 1wt% MWNTs, G' values are equivalent to those recorded for the binary system containing 3wt% MWNTs. In that sample, the spatially-linked structure formed at this clay relative content can thus be reinforced by the addition of a small quantity of carbon nanotubes.

Conclusions

A new type of materials based on a soft polymer matrix, i.e., ethylene vinyl acetate copolymer (EVA) filled with organoclays and multi-walled carbon nanotubes, actually flake-like (2-D) and rod-like (1-D) nanofillers, respectively, have been prepared via direct melt blending. Our study was focused on the characterization of the material morphology and the rheological behavior in order to better understand the previously observed improvement for properties like the increased thermal stability, mechanical performances and flame retardant behavior [12].

This work confirms that exfoliated nanocomposites based on EVA and organoclay can be prepared by direct melt blending. XRD results, when properly interpreted and combined with TEM results, give a much clearer picture of the nanoscale dispersion and overall global dispersion of the filler in the polymer matrix. While long carbon nanotubes are extremely difficult to disperse in EVA, actually forming well-known associated bundles, the simultaneous addition of clay and carbon nanotubes gives a particular morphology: not totally exfoliated clay platelets and partially dispersed long nanotubes with high degree of interconnectivity. The enhanced elastic melt properties, represented by the increase of the storage modulus G', are also particular when the two fillers are simultaneously added. This rheological response is sensitive to the clay amount and related spatially-linked structure, as well as the MWNTs interconnectivity.

Acknowledgments

The authors wish to thank « Région Wallonne » for financial support in the frame of the Nanotechnology project: BINANOCO. This work was partly supported by the Belgian Federal Government Office of Science Policy (SSTC-PAI 5/3). M.A. and Ph.D. are much indebted to both « Région Wallonne » and European Community "FSE and FEDER" for financial support in the frame of Objectif-1 Hainaut : Materia Nova.

[1] Rothon R.; *Particulate-Filled Polymer Composites*, 1st ed. Harlow: Longman Scientific:, 1995.
[2] Giannelis EP; *Adv Mater* 1996; 8: 29-35.
[3] Alexandre M, Dubois Ph; *Mater Sci Eng* 2000; R28: 1-63.
[4] Pinnavaia TJ, Beall GW; *Polymer-clay nanocomposites*, in "Wiley series in polymer science", Chichester : John Wiley and Sons, 2000.
[5] Ijima S; *Nature* 1991; 354: 56-58.
[6] Ajayan PM; *Chem Rev* 1999; 99: 1787-1799.
[7] Baughman RH, Zakhidov AA, de Heer WA; *Science* 2002; 297: 787-792.
[8] Thostenson ET, Ren ZF, Chou TW; *Comp Sci Tech* 2001;61: 1899-1912.
[9] Andrews R, Jacques D, Minot M, Rantell T; *Macromol Mater Eng* 2002; 287: 395-403.
[10] Kashiwagi T, Grulke, E, Hilding J, Harris R, Awad W, Douglas J; *Macromol Rapid Commun* 2002; 23: 761-765.
[11] Pirlot C, Willems I, Fonseca A, BNagy J, Delhalle J; *Adv Eng Mater* 2002; 4: 109-114.
[12] Peeterbroeck S, Alexandre M, BNagy J, Pirlot C, Fonseca A, Moreau N, Philippin G, Delhalle J, Mekhalif Z, Sporken R, Beyer G, Dubois Ph; *Comp Sci Tech*, 2004, in press
[13] Beyer G; *Fire Mater* 2002; 26: 291-293.
[14] Alexandre M, Beyer G, Henrist C, Cloots R, Rulmont A, Jérôme R, Dubois Ph; *Macromol Rapid Commun* 2001; 22: 643-646, and Chem Mater 2001; 13: 3830-3832.
[15] Zanetti M, Camino G, Thomann R, Mülhaupt R; *Polymer* 2001; 42: 4501-4507.
[16] Tang Y, Hu Y, Wang SF, Gui Z, Chen Z, Fan WC; *Polym Deg Stab* 2002; 78: 555-559.
[17] Duquesne S, Jama C, Lebras M, Delobel R, Recourt P, Gloagen JM; *Comp Sci Tech* 2003; 63, 1141-1148.
[18] Riva A, Zanetti M, Braglia M, Camino G, Falqui L; *Polym Degrad Stab* 2002; 77: 299-304.
[19] Lim YT, Ok Park O; *Macromol Rapid Commun* 2000; 21: 231-235.

Macromol. Symp. **2005**, *221*, 127-135

Preparation, Characterisation and Properties of Nanocomposites Based on Epoxy Resins – An Overview

Andreas Hartwig, Monika Sebald, Dirk Pütz, Lisa Aberle*

Fraunhofer-Institut für Fertigungstechnik und Angewandte Materialforschung, Klebtechnik und Oberflächen, Wiener Str. 12, 28359 Bremen, Germany
E-mail: har@ifam.fraunhofer.de

Summary: An overview about the use of nanoparticles in epoxy resins is given. Deaggregated fumed silica, sol gel materials and other spheroidic nanoparticles improved the abrasion resistance and the mechanical properties of filled epoxides, and several properties can be improved by specific particles. Combustion properties, strength and permeation can be improved by organically modified layered silicates. The proper characterisation of the nanocomposites is still an art. TEM analysis of the cured materials and light scattering with 3D cross-correlation for the liquid samples were sufficient methods for the characterisation of filled epoxides.

Keywords: clay; epoxy resin; fumed silica; nanocomposite; particle size

Introduction

Nanofillers used in polymer science are prepared in different ways and have different shapes. Fumed oxides, especially fumed silica, are the most widely used nanofillers. They are characterised by a strongly aggregated and agglomerated structure with more or less round shaped primary particles. With the sol-gel process it is possible to prepare more perfect ball like particles with a minimum of interactions between the particles and very narrow particle size distributions. Furthermore the chemical composition of the particles can be varied in a wide range. Unfortunately the process is difficult to control in big quantities and it is difficult to transfer the particles from the liquid state into a polymer without aggregation.

An other important class of nanofillers used in epoxides are organically modified clay minerals which have a plate-like structure. Most published work[1] is carried out with bentonite as the clay mineral. The organic modification is carried out by the exchange of the sodium ions between the

 DOI: 10.1002/masy.200550313

layers of alumosilicate. Alkyl ammonium salts with at least one long hydrocarbon chain are commonly used as organic ions. The distance between the single layers and the affinity between the particle surface and the polymer increases due to the modification. Both factors are important to produce nanocomposites instead of the conventional microcomposites obtained by using unmodified clay as a filler.

Another case are nanoporous fillers like zeolites or aerogels based on silicon dioxide. Nanoparticles with several different shapes, e.g. needles or cubes, are known but up to now they are at least of minor practical importance.

The properties of the prepared nanocomposites depend strongly on the particle size distribution and mainly on the interactions between the particles themselves and between the particles and the matrix polymer. Therefore, the processing of the materials is at least as important for the properties of the nanocomposites as their chemical composition.

Determination of Particle Size Distribution and Dispersion Stability

Microscopic and scattering techniques are the main methods to analyse the size and distribution of the particles in a matrix material. From these classes of methods transmission electron microscopy (TEM) and laser light scattering are the most important techniques.

For conventional light scattering the samples of interest need to be nearly transparent which in the majority of cases means strong dilution. This requires scrupulous cleanliness of the samples because the tiniest amounts of dust would lead to malfunctional results. 3D-cross-correlation, a new light scattering method, developed in our laboratories, does overcome these disadvantages: turbid samples are accessible in a wide range of concentration.[2, 3] Therefore, small amounts of dust does not influence the results because of the strong scattering signal of the sample. Furthermore, the new technique offers characterisation of a lot of samples in their original state which would be altered or destroyed by dilution. This means that the measured particle size is not influenced by the preparation of the sample. Due to the ability to measure at higher concentrations the method is additionally able to detect the stability of dispersions, which is not possible with conventional light scattering. Figure 1 shows the way of raw data evaluation. Figure 2 illustrates some causes for the change of particle distribution.

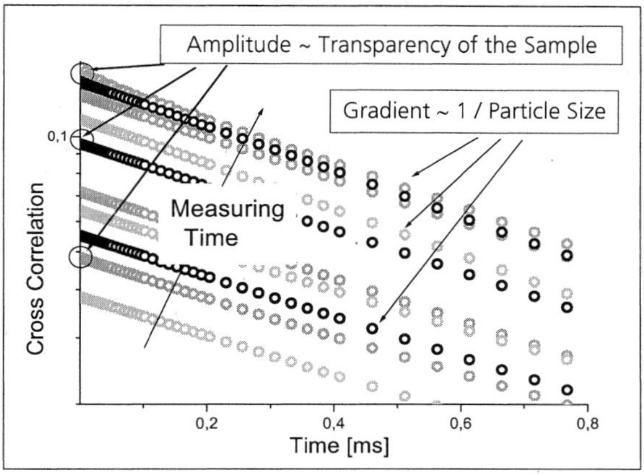

Figure 1. The signal of cross-correlation has two parameters which give information about particle size and stability: With increasing measuring time the sample becomes clearer due to sedimentation. This causes an increase in the amplitude. The gradient gives the particle size which in this case remains unaltered. This means that the sample contains monodisperse particles.

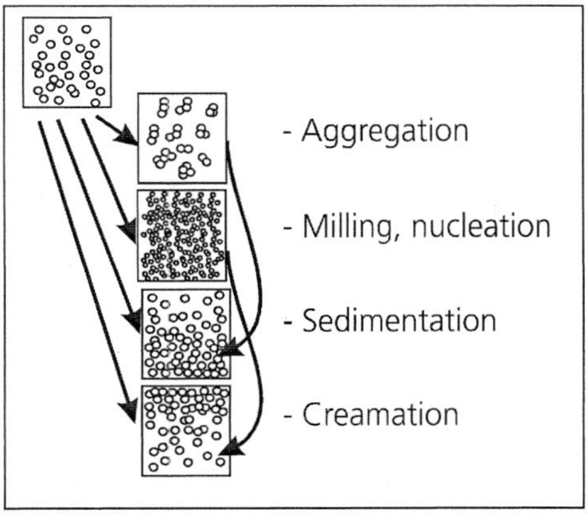

Figure 2. Processes taking place in dispersion leading to a non sufficient stability.

Measured particle size distributions are often a "function of the method used". This might be due to artefacts (e.g. caused by time dependent processes as sedimentation or aggregation as discussed above). Furthermore the different light scattering techniques are sensitive for different ranges of particle sizes. As a rule of thumb instruments based on Fraunhofer and Mie scattering have their strength in characterisation of micro and submicro particles whereas Correlation Spectroscopy is the method of choice for the characterisation of nano particles. It is essential to keep this in mind especially when samples contain particles with a broad or multimodal size distribution. Characterisation of such samples often need an analysation scheme which is a combination of both laser light scattering techniques.[4]

Fumed Oxides and Sol-Gel Particles

Fumed oxides, especially silicon dioxide, have already been used for a long time in polymers, namely in epoxides. It is mainly used to prepare thixotropic formulations and the effects are observed with filler contents in the lower percent range. The achieved changes of the rheological properties are caused by the interaction of the nanoparticles and this is the cause for the restricted maximum amounts of filler in the epoxides.[5] To apply fumed oxides in fields where high amounts of filler or not interacting particles are required, a deaggregation is necessary.

Coatings with high abrasion resistance are required e.g. for parquet, automotive clear coats or the improvement of the scratch resistance of parts made out of transparent polymers like polycarbonate. In recent times the abrasion resistance of coatings was improved by the modification with nanoparticles.[6] The nanoparticles were prepared by the so called sol-gel-process. As the particles are prepared in situ, the applicability is restricted to a limited number of polymers as base material and tailor made products are required for each application. Furthermore the complete filler is synthesised by the hydrolysis of silanes, which is a quite complicated process. To obtain similar results with fumed oxides, a deaggregation and high filler loadings are required. This is obtained by a combination of surface silylation of fumed silica and mechanical processing. In contrast to the sol-gel synthesis of nanoparticles it is not possible to dispense single particles in the organic matrix. TEM examinations (Figure 3) as well as light scattering show that very small agglomerates are formed. The size of theses agglomerates is typically below 100 nm, which is also the reason for the high transparency of the filled polymers. Even at loadings of 50 wt-% the materials are as transparent as the non filled polymers. As the interaction between the particles is reduced, due to the modification, the rheology of the resulting nanocomposites even at filler contents as high as 30 wt-% is only slightly influenced compared to the unfilled resins. Formulations containing the modified nanoparticles and curable resins are applicable as coatings with reduced abrasion. Compared to the non-filled materials the abrasion measured by the Taber abrasion test is reduced by a factor of 2 to 5 for coatings with a filler content of 25 wt-% (Figure 4). The abrasion of an epoxide based coating was reduced from 39 mg to 9 mg (size of the primary particles 12 nm) and 16 mg (size of the primary particles 40 nm) respectively. The much lower abrasion of the material with the 12 nm particles compared to that with the 40 nm particles shows that a real nano-effect is the cause for the improvement.[7]

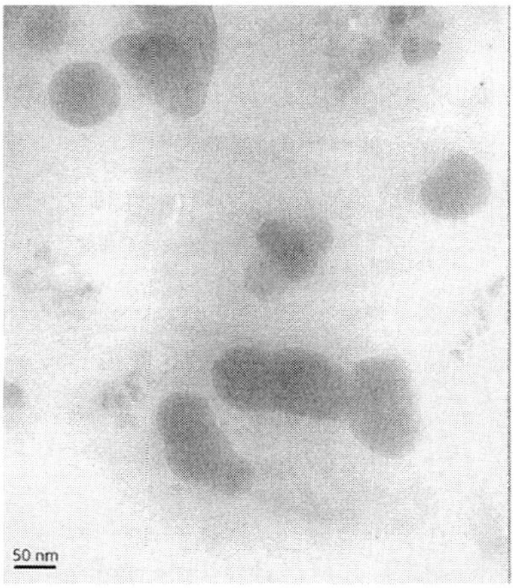

Figure 3. TEM image of a sample with 20 wt-% filler; size of primary particles 40 nm.

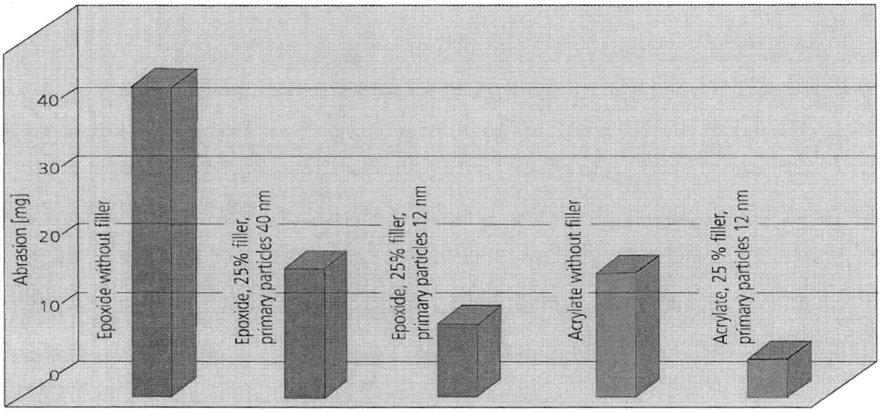

Figure 4. Abrasion properties of nanocomposites

By the incorporation of sol-gel based silica nanoparticles in an epoxy resin the expected increase of the modulus was observed. But interestingly the toughness increases, too. This is in contradiction with the experience made with microparticles in polymer matrices. The bending strength of a glass fibre reinforced epoxide could also be improved with the sol-gel particles. The increase is only observed for particle concentrations up to 10 wt-%, indicating that an interaction between the particles is not the cause for the effect but an interaction between the particle surface and the base polymer.[8] An improvement of the adhesion strength of epoxy resins was also observed for the deaggregated fumed oxides described above.

If additional effects are required beside the improvement of the mechanical properties, other nanoparticles than silicon dioxide are required. An example are indium tin oxide nanoparticles which are suitable to prepare transparent and electrically conductive nanocomposites. These nanocomposites are appropriate heat shielding coatings or can be used for antistatic purposes. Another example are superparamagnetic iron oxide particles. Nanocomposites containing such particles can be heated in a high frequency field, e.g. for curing of epoxides. Conventional electrically conductive adhesives contain silver flakes. Alternatively nanosized silver particles forming aggregates and agglomerates can be used. With this kind of filler the required amount to obtain electrical conductivity could be reduced from about 70 to 55 wt-% and the adhesive connections prepared with filled epoxides have a higher elongation at break and a higher strength if the silver nanopowder is used. At low concentration of silver nanoparticles the polymers, especially the epoxide based adhesives, have a bactericide or bacteriostatic surface depending on the amount of silver used and the polymer composition. The materials filled with micrometer sized silver particles does not show this effect.

Layered Silicates

Layered silicates, namely organically modified bentonites, are used to improve a number of properties of the respective matrix polymers.[9] An optimum improvement is obtained with loadings in the range of 5 wt-% or even lower amounts of bentonite. The most prominent examples are the improvement of the mechanical strength without lowering or even elevating the toughness. The reduction of the permeability of packaging materials, e.g. for the use as bottles or foils, is also an important field of activity. Both kind of effects - reduced permeability and improved mechanical properties - are also important for epoxy resins. Such epoxy resins are for example applicable as adhesives with improved strength and durability. The third important field for the application of organobentonites in polymers is the improvement of the combustion properties. This is shown in more detail for epoxy resins modified with 5 wt-% organobentonite.[10]

A cyloaliphatic and a bisphenol A diglycidylether based epoxy resin were filled with 5 wt-% organobentonite with different organic modifications. The combustion properties of the samples were tested with a horizontal burner test and the reduction of the combustion rate compared to the pure epoxy resin was determined. The results are shown in Figure 5. Although the absolute combustion rate of the cycloaliphatic epoxy resin is much higher than that of the aromatic system, the relative improvement is the same for both systems as long as the same organic modification of the bentonites is used. By a systematic variation of the organic modification of the bentonite it is possible to reduce the combustion rate significantly. The combustion rate of the cycloaliphatic epoxide could be reduced to 12 % of that of the non-filled polymer.

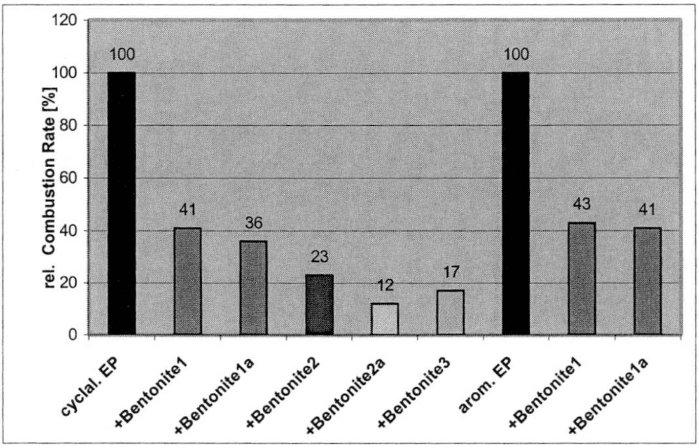

Figure 5. Relative combustion rate of epoxides filled with 5 wt-% organobentonite with different modification.

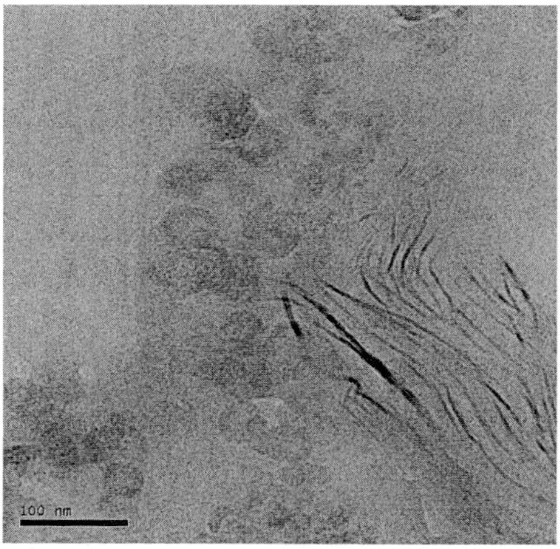

Figure 6. TEM image of the intersection between incinerated and intact area of a nanocomposite consisting of a cycloaliphatic epoxy resin and 5 wt.-% organobentonite (Image taken from the cover page of Macromol. Chem. Phys. **204** (2003), issue 18).

The proposed flame retardation mechanism of bentonite based nanocomposites is the formation of a char like barrier layer. The intersection between the incinerated and the intact area of the samples was examined by TEM (Figure 6). It can be seen that the particles are destroyed during combustion and that the formed barrier layer consists of particles.

Conclusions

A number of properties of epoxy resins can be adjusted by the addition of nanoparticles. To observe the desired effects a reproducible and homogeneous distribution of the particles is important. This is only possible if the particle surface has a high enough affinity to the epoxy resin. If the affinity between the particles is higher compared to the affinity between particles and polymer reaggregation takes place or no homogeneous distribution of the particles is observed. Interestingly the effects are often observed with already small amounts of nanoparticles, indicating the importance of the interaction between particle surface and polymer. Furthermore the use of nanoparticles seems to be a way to combine properties which are in combination not possible with conventional materials, e.g. increasing strength combined with increasing toughness.

Acknowledgements

Financial support of the presented work by the State of Bremen within the AMST program and the German Ministry of Education and Research (BMBF) is gratefully acknowledged.

[1] Z. Wang, J. Massam, T.J. Pinnavaia, in: *"Polymer-Clay Nanocomposites"*, T.J. Pinnavaia, G.W. Beall, Eds., J. Wiley & Sons, Chichester 2000, p. 127ff
[2] L.B. Aberle, W. Staude, O.-D. Hennemann, *Phys. Chem. Chem. Phys.* **1999**, 1, 3917.
[3] L.B. Aberle, P. Hülstede, S. Wiegand, W. Schröer, W. Staude, *Applied Optics* **1998**, 37, 6511.
[4] R. Buttgereit, T. Roths, J. Honerkamp, L. Aberle, *Phys. Rev. E* **2001**, 64, 1.
[5] Degussa, Technical Bulletin AEROSIL No. 27, AEROSIL for Solvent-Free Epoxy Resins, **2001.**
[6] C. Roscher, *Europ. Coat. J.*, **04/2003**, 138.
[7] A. Hartwig, M. Sebald, *Macromol. Chem. Phys.* **2003**, 204 (2), F51.
[8] S. Sprenger, C. Eger, A. Kinloch, A.C. Taylor, J.H. Lee, D. Egan, *Adhäsion - kleben & dichten* **03/2003**, 24.
[9] A. Alexandre, P. Dubois, *Mat. Sci. Eng.* **2000**, 28, 1.
[10] A. Hartwig, D. Pütz, B. Schartel, M. Bartholmai, M. Wendschuh-Josties, *Macromol. Chem. Phys.* **2003**, 204, 2247.

Reprocessable and Easy Processing Silica Filled Brominated Isobutylene para Methyl Styrene (BIMS)

Wai Keung Wong

ExxonMobil Chemical Europe, Hermeslaan 2, B1831 Machelen, Belgium
E-mail: john.k.wong@exxonmobil.com

Summary: Without the use of any curatives, silica filled BIMS compounds can achieve tensile strength and bound rubber level close to conventional crosslinked elastomer compounds. This outstanding tensile performance of silica filled BIMS compounds results from the strong interaction between BIMS polymer and silica filler. Silica filled BIMS compounds can be reprocessed and still retain their high tensile strength performance. The good compatibility between BIMS and silica also leads to better filler dispersion and inhibition of filler-filler interaction. This in turn leads to the lower processing viscosity observed. We speculate that BIMS can interact with silica via nucleophilic substitution reaction between benzylic bromide of the polymer and surface silanol group of silica.

Keywords: BIMS; interaction; processing; reprocessability; silica

Introduction

In industrial applications, elastomers are typically compounded with various particulate fillers such as carbon black, silica, silicate, clay etc. [1-6] Most of these fillers act as reinforcement to the polymer matrix and can also reduce compound cost.

In comparison to carbon black, silica filled elastomers can provide certain performance advantage such as tear strength. However, silica filler is more difficult to disperse in non-polar elastomers and requires special silane coupling agent and mixing conditions for effective incorporation. Despite the above, there is a recent surge of interest in using silica as filler. This is because silica filled elastomers can provide low rolling resistance tread composition and offer a better balance of traction and treadwear in comparison to carbon black filled elastomer [7].

Silica generally has a higher tendency than carbon black to form filler-filler interaction in polymer matrix, leading to higher processing viscosity [4].

 DOI: 10.1002/masy.200550314

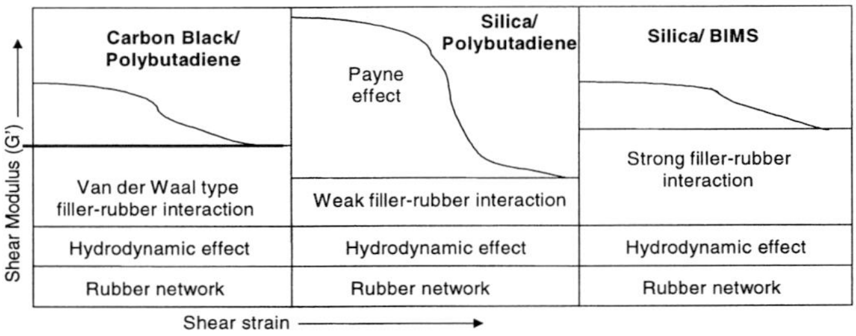

Figure 1. Schematic diagram showing the contribution factors of shear modulus in various filled compounds.

Figure 1 shows that silica filled polybutadiene has dramatically higher shear modulus (G') at low strain due to the presence of strong filler-filler interactions. This is known as the Payne effect [8, 9]. At high strain amplitude, the filler-filler linkage breaks down by the dynamic straining and as a result a decrease in shear modulus is observed [10]. Carbon black filled polybutadiene and silica filled BIMS do not exhibit significant Payne effect indicating a reduction in filler-filler interaction.

There is little interaction between surface silanol groups of silica and non-polar polymer matrices. In order to overcome these drawbacks, mercapto silanes such as triethoxysilyl propyl tetrasulfane (Si69) that can react with silanol groups of silica (through ethoxyl) as well as the double bond in unsaturated elastomers (through tetrasulfane) have been developed and used by the industry [11]. These mercapto silanes also promote dispersion of silica in elastomers and reduce filler-filler interaction, leading to improved processability and better physical and dynamic performances of filled elastomer.

Experimental

All the compounds were mixed in the same manner using a 300cc Haake Rheomix with a minimum of 3 minutes mixing time at a constant temperature of 150°C to ensure optimal silanization reaction. Finalization was carried out using an open mill set at 60°C.

The bound rubber tests were carried out by first dissolving the samples in toluene at room temperature for 1 week. The solution was then filtered and any sample remaining on the filter was allowed to dry under vacuum at 80°C for 3 days or until the weight becomes constant. The % bound rubber of sample is calculated using the equation:

Bound rubber (%) = [(wt of dried sample remain on filter - wt of filler in the sample) X 100] / (wt of polymer in the sample) (1)

Results and Discussion

Figure 2 shows the Mooney viscosity change with temperature of BIMS (Exxpro 3745) (note: Exxpro is the trade name of BIMS) and polybutadiene (Budene 1207, PD) as well as N660 black or high dispersion silica (Zeopol 8745) filled compound. All compound contains 50 phr filler (when present), 20 phr of oil, 3 phr of Si69 (when present) and curing agents. It can be observed that silica filled polybutadiene exhibits much higher Mooney viscosity than silica filled Exxpo 3745 compound. This is attributed to the extensive silica network formation in the silica filled polybutadiene compound [12, 13].

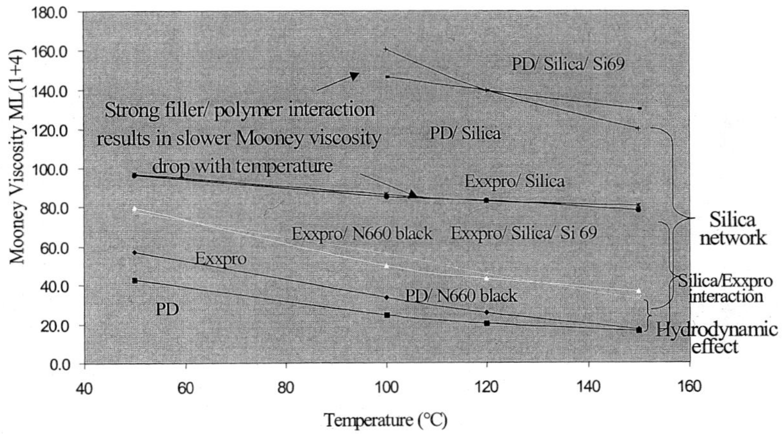

Figure 2. Mooney viscosity versus temperature plot of carbon black and silica filled model compounds.

It can also be observed that the rate of decrease of Mooney viscosity with temperature for silica filled polybutadiene is significantly less when Si69 is present. This is because Si69 promotes strong interactions between silica and polybutadiene and therefore rendering the compound more resistant to flow. The rate of decrease of Mooney viscosity with temperature for silica filled Exxpro 3745, with or without Si69, is similar to that of silica filled polybutadiene that contains Si69. This indicates that BIMS can promote good interaction with silica and minimize silica network formation even without silane.

The amount of bound rubber of silica (Zeopol 8745) and carbon black (FEF N220) filled model compounds that consist of only polymer, filler and in some cases Si69 have been determined and shown in Figure 3.

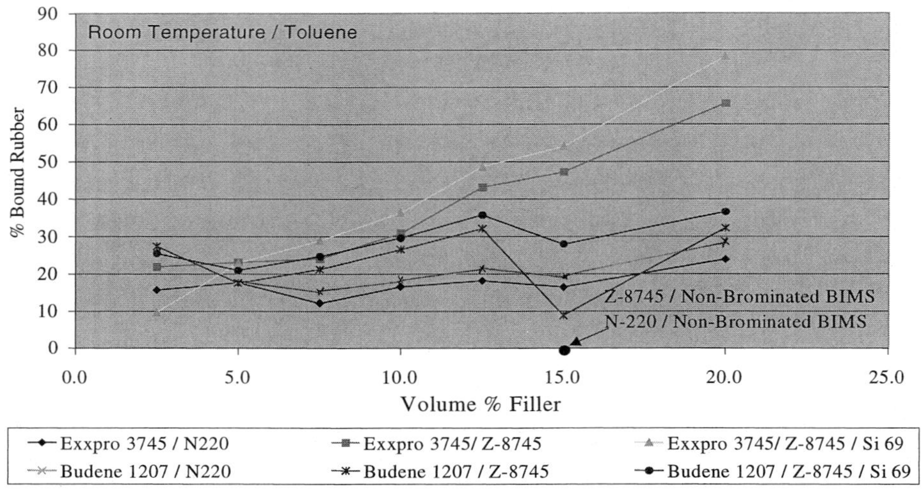

Figure 3. Bound rubber level of silica and carbon black filled model compound.

It can be observed that silica has much higher affinity to BIMS than to polybutadiene while carbon black exhibits similar affinity to both polymers. It can also be concluded that benzylic bromide is responsible for the high bound rubber level observed in silica filled BIMS. Non-brominated BIMS has zero bound rubber for both silica and carbon black filled samples. The much higher bound rubber level of silica filled BIMS compound in comparison to the carbon black filled one further confirms the existence of good interactions between BIMS and silica.

Table 1 compares the tensile performances of silica filled polybutadiene (PD) and Exxpro 3745 to that of crosslinked elastomer compounds.

Table 1. Mooney viscosity and tensile properties of silica filled compounds.

Tensile bar molding conditions		**180°C for 20 minutes**		
	Mooney Viscosity ML 1+4 (100°C)	100% Modulus (MPa)	Max. Tensile Strength (MPa)	Elongation at break (%)
PD/ silica	>200	1.6	3.0	270
PD/ silica/ Si69	>200	1.5	2.4	270
Exxpro/ silica	139	2.8	4.4	325
Exxpro/ silica/ Si69	130	3.6	8.1	295
Crosslinked EPDM hose compound	-	3.2	10.5	400
Crosslinked Bromobutyl innerliner compound	-	0.8	9.1	895

All the silica filled, non-crosslinked polybutadiene and BIMS compounds contain 20 volume % filler and 3.3 phr of Si69 (if present) without any oil or curing agent. It can be observed that the silica filled BIMS compound has much lower Mooney viscosity in comparison to the silica filled polybutadiene. Furthermore, tensile properties similar to crosslinked elastomer compounds can be achieved for the silica filled BIMS compound. It can also be observed that the use of Si69 further enhances tensile properties of silica filled BIMS but has no effect on silica filled polybutadiene. This indicates that the incorporation of Si69 leads to further improvemnt of silica reinforcement of BIMS. Si69, bound to the silica surfaces, promotes interaction between unsaturated polymer and silica via reaction of the tretasulfane with unsaturations [4]. Since BIMS polymer contains no unsaturation and crosslinking is via Friedel-Crafts mechanism [14] rather than via radical or ionic reactions that involves sulphur, the mercapto functional group of Si69 is unlikely to be involved in the interaction between silica and BIMS. It has also been shown that sulphur has no effect on the crosslinking of BIMS [15]. The enhancement of tensile properties observed due to the presence of Si69 most likely results from a nucleophilic substitution reaction between benzylic bromide of BIMS and surface silanol group of silica:

$$\underset{H}{Si\text{-}O^-} + Br\text{-}CH_2\text{-}PH \rightarrow \underset{H}{Si\text{-}O}\text{-}CH_2\text{-}PH + Br^- ;$$

where PH stands for phenyl group.

This type of well known displacement reaction is best exemplified by the conversion of an alkyl halide into an alcohol by aqueous base:

$$HO^- + R\text{-}Hal \rightarrow HO\text{-}R + Hal^- ;$$

where Hal stands for halogen such as Br, Cl, I.

Therefore the role of the silane coupling agent (Si69) is simply to further improve silica dispersion and therefore exposes more filler surfaces for interacting with BIMS. This is illustrated in Table 2 which shows that a variety of silane coupling agents, irrespective of their functional group, can further enhance the physical performances of silica filled BIMS (Exxpro 3745).

Table 2. Effect of silane coupling agents on tensile properties of silica filled Exxpro 3745 compounds.

Tensile bar molding conditions		180°C for 20 minutes		
All compounds contain 20 vol % filler without any curing agent or oil	Mooney Viscosity ML 1+4 (100°C)	100% Modulus (MPa)	Max. Tensile Strength (MPa)	Elongation at break (%)
No silane				
No silane	139	2.8	4.4	325
Alkyl silane				
PTMO (0.5)	139	2.9	5.6	350
PTEO (1.2)	138	2.5	5.7	400
A1630 (1.6)	>200	3.0	7.3	280
Mercapto silane				
Si69 (3.3)	130	3.6	8.1	295
MTMO (1.5)	125	3.3	5.1	320
Methacrylate silane				
Y11878 (1.6)	137	2.5	5.1	415
Amino silane				
AMMO (2.2)	141	3.6	5.2	260

The silane coupling agents listed in Table 2 are:

PTMO: n-propyl trimethoxysilane

PTEO : n propyl triethoxysilane

A1630: n-methyl trimethoxysilane

Si69: Bis(3-triethoxysilylpropyl)tetrasulfane

MTMO: 3-mercaptropropyl trimethoxysilane

Y11878: γ-methacryloxypropyl triethoxysilane

AMMO: 3-aminopropyl trimethoxysilane

Note: Numbers in brackets represent parts per hundred (phr) silane relative to the polymer

The above tends to suggest that the high tensile strength performances and high bound rubber level of silica filled BIMS is achieved mainly via strong interaction between filler and polymer. Table 3 shows that the silica-BIMS interaction does not lead to a three-dimensional crosslink network structure formation in the compound. Unlike conventional crosslinked rubber, silica filled BIMS can be remilled for 5 minutes using an open mill set at 120°C and then remolded under the same conditions (180°C for 20 minutes) and still retain their high tensile properties.

Table 3. Mooney viscosity and tensile properties of silica filled Exxpro 3745 compounds before and after remoulding.

Tensile bar molding conditions		180°C for 20 minutes		
All compounds contain 20 vol % filler without any curing agent or oil	Mooney Viscosity ML 1+4 (100°C)	100% Modulus (MPa)	Max. Tensile Strength (MPa)	Elongation at break (%)
Exxpro/ Silica	139	2.8	4.4	325
Remilled/ remolded	146	-	-	-
Exxpro/ Silica/ Si69 (1.5)	139	3.3	8.2	315
Remilled/ remolded	147	2.6	7.5	360
Exxpro/ Silica/ Si69 (3.3)	130	3.6	8.1	295
Remilled/ remolded	159	2.8	7.9	270

Conclusions

In this paper it is demonstrated that compounds of silica filled **B**rominated **I**sobutylene para**M**ethyl **S**tyrene (BIMS) is easier to process, as indicated by the lower Mooney viscosity, than similar compounds based on unsaturated polymers such as polybutadiene. Furthermore, the use of unconventional alkyl silanes can further improve processability as well as physical properties of silica filled BIMS while the same silanes have no effect on silica filled polybutadiene. Without the use of any curatives, silica filled BIMS compounds can achieve tensile strength and bound rubber level close to conventional crosslinked elastomer compounds. This outstanding tensile performance of silica filled BIMS compounds results from the strong interaction between BIMS polymer and silica filler instead of the crosslinking of polymer matrix. Furthermore, silica filled BIMS compounds that exhibit tensile strength of 8 MPa and elongation to break of around 300%, as well as bound rubber level of 80%, can be reprocessed by roll milling and still retain their physical performance after reprocessing. The good compatibility between BIMS and silica also leads to better filler dispersion and inhibition of filler-filler interaction. This in turn leads to lower processing viscosity. We speculate that BIMS can interact with silica via substitution reaction between benzylic bromide of the polymer and surface silanol group of silica.

[1] G. Kraus, ed. *Reinforcement of Elastomers,* Wiley-Interscience, New York, **1965**.
[2] G. Kraus, *Rubber Chem. Technol.* **1965,** 38, 1070.
[3] G. Kraus, *Agnew. Makromol. Chem.* **1977** 60/61, 215.
[4] S. Wolff, *Rubber Chem. Technol.* **1996,** 69 (3), 325-346.
[5] J-B Donnet, *Rubber Chem. Technol.* **1998,** 71 (3), 323-341.
[6] M. Gerspacher, C.P.O'Farrell, H.H. Yang and L. Nikiel, *Tire Technology Int.*, **April 1999**, 27-33.
[7] J.T. Byers, *Rubber Chem. Technol.* **2002**, 75(3), 527-547.
[8] A.R. Payne, *J. Polym. Sci.* **1962**, 6, 57.
[9] A.R. Payne and R.E. Whittaker, *Rubber Chem. Technol*, **1971**, 44, 440.
[10] M.J. Wang, *Rubber Chem. Technol.* **1999**, 72 (2), 430-448.
[11] S. Wolff, *European Rubber Journal*, **January 1994**, 16-19.
[12] M.F. Tse, H.C. Wang, J.E. Rogers, *Rubber World,* **June 1997**, 39-45.
[13] W.K. Wong, G. Ourieva, M.F. Tse, H. C. Wang, *Presentation at Eurofiller 2001*, **July 9-12**, LODZ.
[14] R.Bielski, J.M.J. Frechet, J.V. Fusco, K.W. Powers, H.C. Wang, *Journal Poly. Sci, Part A: Poly. Chem.***1993,** 31, 755-762.
[15] Unpublished data.

Effect of Unmodified and Surface Treated Fumed Silica on the Polymerization of HEMA

Valentin Tertykh, Yulia Bolbukh, Victor Yanishpolskii*

Institute of Surface Chemistry of the National Academy of Sciences of Ukraine, 17 Gen. Naumov Str., 03164 Kyiv, Ukraine
E-mail: tertykh@public.ua.net

Summary: Thermal polymerization of 2-hydroxyethylmethacrylate (HEMA) and radical polymerization initiated by an acrylamid complex of cobalt nitrate in the presence of unmodified and surface treated silicas have been investigated. It has been found that the introduction of fillers, especially silicas with grafted silicon hydride groups, makes an essential effect on the rate and degree of the HEMA polymerization.

Keywords: cross-linking; fillers; hydride-containing silicas; 2-hydroxyethyl-methacrylate (HEMA); radical polymerization

Introduction

In many cases essential strengthening of a filled polymeric system can be achieved when active sites of modified surface of a silica filler are able to interact chemically with functional groups of polymer macromolecules, which results in chemical cross-linking. Such a chemical cross-linking was realized for plastic binders and filled polymeric systems with the help of finishing agents with active functional groups for surface modification of glass fibers and disperse silicas.[1, 2] Among modifying reagents, organosilanes with amino-, methacryl-, epoxy- or vinyl groups have much use, and their filler/polymeric matrix interface structures have been widely studied. Studies of behavior of silicon hydride-containing silicas as active fillers of polymeric systems are practically absent. Recently, systematic study of the hydrosilylation reaction of functional olefins by silicon hydride groups attached to silica surface under the action of catalysts and heat treatment were carried out.[3-5] The proceeding of surface chemical reaction involving the grafted surface ≡SiH groups and olefin molecules is effected according to the following general scheme:

 DOI: 10.1002/masy.200550315

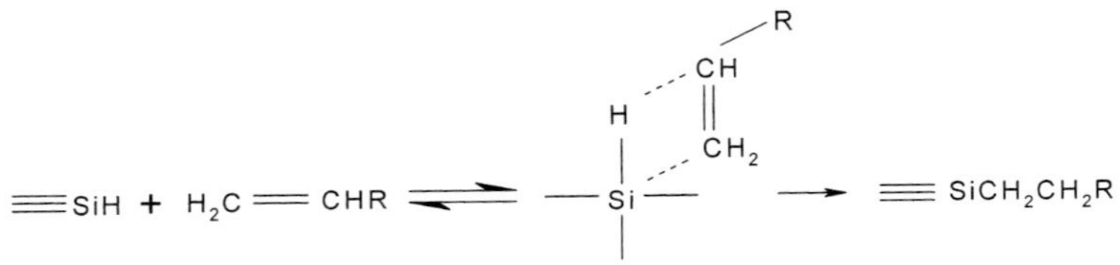

where *R* is a suitable organic radical. In line with these data, silicas with the grafted silicon hydride groups on the surface layer are of particular importance as chemically active fillers of the olefin-containing polymers.

Investigation of the interaction of silicon hydride-containing silicas with derivatives of methacrylic acid is of special interest. The aim of this work is to study the role of hydrosilylation reaction on the thermally induced self-polymerization and initiated polymerization of HEMA to obtain data concerning the effect of surface groups on the polymerization.

Experimental

Hydride-containing silicas were prepared by the reaction of triethoxysilane (Kremnepolymer, Zaporizhya) with fumed silica (Oriana, Kalush, 300 m^2/g) at room temperature in ethyl alcohol under stirring for 24 h. Then the modified silicas were dried at 100 °C.

2-Hydroxyethylmetacrylate (Fluka) was used without any additional purification.

Acrylamide (AAm) complex of cobalt nitrate of composition $4AAm\cdot Co(NO_3)_2\cdot 2H_2O$ was synthesized in accordance with the recommendations of Savost'yanov et al.[6]

Thermal HEMA polymerization was conducted in an open reactor at 106 °C. The polymerization was carried out in the presence of unmodified fumed silica or hydride-containing silicas with concentration of bound ≡SiH groups of 0.25 and 0.64 mmol/g. Filler content was equal to 16 wt.% of composite.

Radical HEMA polymerization initiated by acrylamide complex of cobalt nitrate was performed at contents of the initiator of 2, 5, 10, 20 wt.% at 96 °C and at the initiator–monomer mass ratio of 1:10 at 96, 106, 116 °C. The polymerization of filled systems was carried out for 10 wt.% of the initiator and content of the silica filler of 16 wt.%. Prior to the polymerization the initiator and filler were carefully mixed and then added to the monomer.

The concentration of C=C groups in the composite system was determined by a titrimetric (bromide-bromate) method by analyzing water extracts of the appropriate samples after separation of the polymer.

Results and Discussion

Thermal HEMA polymerization

When studying kinetics of the thermal HEMA polymerization at 106 °C (Fig. 1, *a*) we observed a considerable scatter between individual determinations of C=C group concentrations (*C*, mol/L). As in the case of other vinyl monomers[7], it may be related to formation of the low-molecular adducts which affect the polymerization rate. Kinetic curves have a plateau-like section corresponding to an induction period with a subsequent acceleration of the polymerization process. The monomer contains hydroquinone monomethyl ether as a stabilizer, which may at least influence the start of the reaction and it is possibly the reason for the induction period.

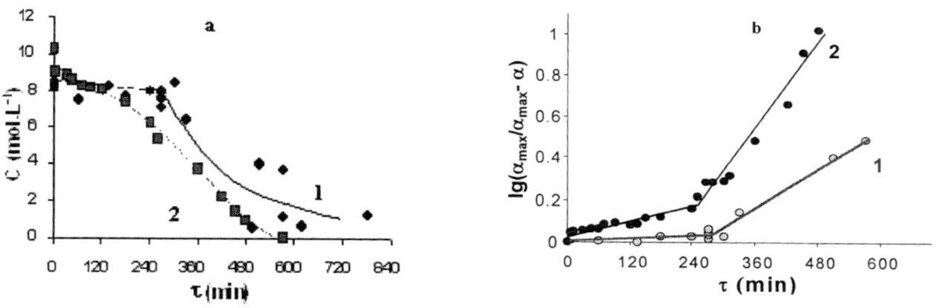

Figure 1. Kinetics of HEMA polymerization at 106 °C (*a*) and anamorphosis of the kinetic curves (*b*) in the absence (*1*) and in the presence (*2*) of acrylamide complex of cobalt nitrate.

The polymerization rate constant calculated[8,9] from the equation:

$$k = \frac{2.303}{\tau} \cdot \lg \frac{C_o}{C}$$

where C_o is the initial concentration of vinyl groups in the monomer and C is the concentration of C=C groups at time τ, was found to be equal to $5.55 \cdot 10^{-5}$ sec^{-1}. The construction of the logarithmic anamorphosis of kinetics curves[10] $\log\alpha_{max}/(\alpha_{max}-\alpha) = f(\tau)$, where $\alpha = (C_o-C)/C_o$ is the degree of polymerization at time τ and α_{max} is the maximum degree of the monomer transformation, allowed us to discriminate more distinctly both sections of the kinetic curves (Fig. 1, *b*). As it is seen, the introduction of the initiator leads to stabilization of the polymerization process.

Effect of fillers on thermal HEMA polymerization

The study of thermal polymerization of a system on the basis of the monomer and fumed silica (Fig. 2, curve *1*) showed that introduction of the filler resulted in an increase in the polymerization degree. At the same time, it made only a slight effect on the polymerization rate in comparison with pure monomer and led to formation of impure structures. This can be explained by the strong monomer adsorption on the filler surface since the adsorption inhibits formation of the cross-linked structures.[11]

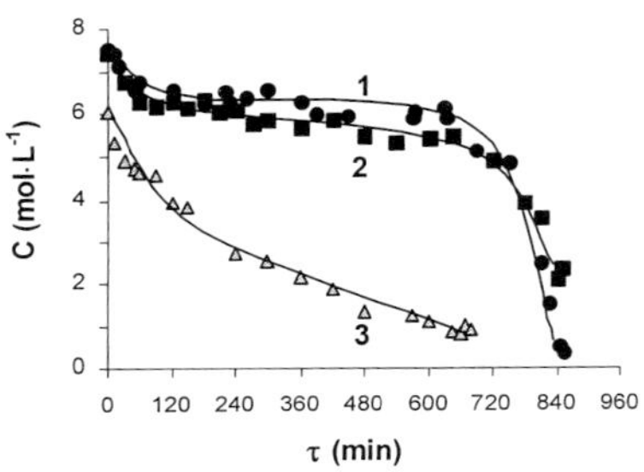

Figure 2. Kinetics of the thermal HEMA polymerization in the presence of as-received fumed silica (*1*) and modified silicas at ≡SiH group concentrations of 0.25 (*2*) and 0.64 mmol/g (*3*) at 106 °C.

Table 1. Time of polymerization and degree of conversion of C=C groups in products of the thermal and initiated HEMA polymerization in the presence of silica fillers at 106 °C.

Composition*	*a*	*b*	*c*	*d*	*e*
Time, h	14	11	6.5	5.5	4.5
$100\cdot(C_o-C)/C_o$, %	69	85	71	74	86

**a*, *b*: HEMA + modified silica with C_{SiH}=0.25 and 0.64 mmol/g respectively;
c: HEMA + acrylamide complex of cobalt nitrate+silica;
d, *e*: HEMA + acrylamide complex of cobalt nitrate+modified silica with C_{SiH}=0.25 and 0.64 mmol/g respectively.

Introduction of hydride-containing silica into the monomer leads to an increase in the polymerization degree. Thus, the concentrations of C=C groups in the resulting composites filled with modified silicas at ≡SiH group concentrations of 0.25 and 0.64 mmol/g were equal to 2.31 and 0.91 mol/L respectively. As the concentration of ≡SiH groups increases, the time of the polymerization decreases from 14 down to 11 h (Table 1).

Accordingly, under the reaction conditions the hydride groups attached to silica surface interact with vinyl groups of the monomer and promote the HEMA polymerization. On the basis of the experimentally observed dependencies it is possible to assume that hydrosilylation of monomer molecules by surface silicon hydride groups results in immobilization of macromolecules, which leads to an increase in the rate and degree of polymerization (Figure 2, curve *3*).

Initiated HEMA polymerization

According to the results achieved, with increasing temperature the polymerization rate decreases, which may be attributed to an influence of the medium viscosity.

Figure 1 shows the effect of acrylamide complex of cobalt nitrate on kinetics of the initiated HEMA polymerization in the comparison with the thermal polymerization in the absence of the initiator at the same temperature. In the presence of the initiator the curve section corresponding to the induction period of the process is diminished and an increase in the polymerization rate is observed.

Because the acrylamide complex of cobalt nitrate is rarely applied as an initiator of the olefin polymerization, the order of the initiation reaction was determined experimentally. The initiation order was equal to 0.53 ± 0.07, which is in agreement with literature data for such widely used radical polymerization initiators as benzoyl peroxide or azobisisobutyronitrile.[12] The inference about the HEMA polymerization by the radical mechanism in the presence of acrylamide complex of cobalt nitrate is corroborated by the linear dependence of the polymerization time on the square root of the initiator concentration.

Initiated polymerization in a filled system

The literature data about an effect of filler surface on initiated polymerization processes are contradictory. In accordance with the work of Tsubokawa et al.[13] the introduction of fumed silica into methylmethacrylate does not influence the degree of polymerization initiated by benzoyl peroxide. However, according to the data of Almazova et al.[14] the fumed silica is an active filler accelerating the formation of the cross-linked structure. Our experimental data give evidence to the fact that the introduction of fumed silica leads to a decrease in the polymerization time from 9.0 down to 6.5 h as compared to the corresponding system without any filler.

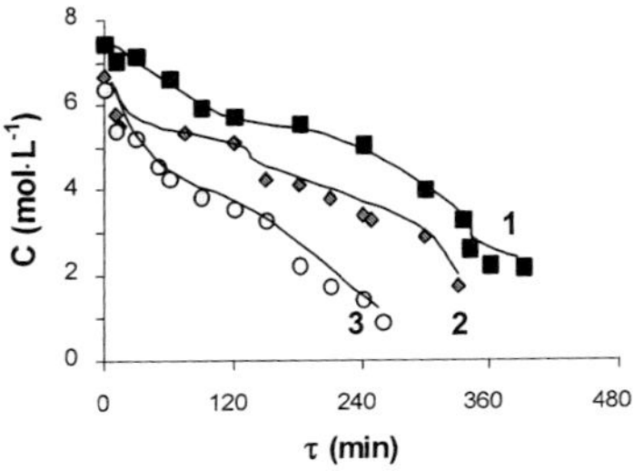

Figure 3. Kinetics of the initiated HEMA polymerization in the presence of the as-received fumed silica (*1*) and modified silicas at ≡SiH group concentrations of 0.25 (*2*) and 0.64 mmol/g (*3*) at 106 °C.

Introduction of silicon hydride-containing silicas promotes an increase in rate and degree of the HEMA polymerization, which may assigned to the presence of additional active surface sites reacting with polymer macromolecules (Figure 3). As an example, the introduction of the hydride-containing silica at a concentration of the grafted ≡SiH groups of 0.25 mmol/g decreases the polymerization time to 5.5 h at the residual C=C groups content of about 1.71 mol/L. If the concentration of ≡SiH groups in the modified filler is 0.64 mmol/g, the polymerization time and the C=C groups content are 4.5 h and 0.89 mol/L, respectively. The presence of silicon hydride groups on the filler surface practically eliminates an induction period of the polymerization reaction.

Conclusions

The results achieved give evidence for the fact that the introduction of the hydride-containing silicas in the case of the thermally induced self-polymerization or initiated polymerization of HEMA provides cross-linking of composite systems due to the reaction between C=C groups of the monomer and ≡SiH groups attached to the filler surface. During the polymerization initiated by the cobalt complex the process of grafting of HEMA molecules onto the filler surface may be affected by grafting acrylamide complexes of cobalt nitrate or by adding of HEMA molecules. The presence of acrylamide ligands contributes to immobilization of HEMA macromolecules and to formation of cross-linked structures. As the ≡SiH group concentration on the silica surface increases, the rate and degree of polymerization become higher.

[1] E.P. Plueddemann, in: "*Composite Materials*", v.6, L.J. Broutman and R.H. Krock (Eds.), Academic Press, New York and London 1974, p. 181.
[2] S. Hayashi, K. Fujiki, and N. Tsubokawa, *React. Funct. Polym.* **2000**, *46,* 193.
[3] V.A. Tertykh and L.A. Belyakova, in: "*Adsorption on New and Modified Inorganic Sorbents*", A. Dabrowski and V.A. Tertykh (Eds.), Elsevier, Amsterdam 1996, p. 147.
[4] V.A. Tertykh, V.V. Yanishpolskii, L.V. Bereza, J.J. Pesek, and M.T. Matyska, in: "*Chemistry, Physics and Technology of Surfaces*", Issues 4-6, A.A. Chuiko (Ed.), KM Academia, Kiev 2001, p. 69.
[5] J.J. Pesek, J.E. Sandoval, D. Chu and E. Jonsson, in: "*Chemically Modified Oxide Surfaces*", H.A. Mottola and J.R. Steinmetz (Eds.), Elsevier, Amsterdam 1992, p. 57.
[6] V.S. Savost'yanov, A.D. Pomogailo, B.S. Selenova, D.A. Kritskaya, and A.N. Ponomarev, *Izv. AN SSSR, Ser. khim.* **1990**, *4*, 768.
[7] J. Rabek, "*Experimental Methods in Polymer Chemistry*", J. Wiley & Sons, New York 1980.
[8] I.V. Berezin and A.A. Klesov, "*Practical Course of Chemical and Enzymatic Kinetics*", Moscow University, Moscow 1976.
[9] G. Odian, "*Principles of Polymerization*", J. Wiley & Sons, New York 1981.
[10] S.I. Omelchenko, "*Complex Oligoethers and Polymers on its Basis*", Naukova Dumka, Kiev 1976.
[11] Y.S. Lipatov, *Prog. Polym. Sci.* **2002**, *27*, 1721.
[12] C.H. Bamford, A.D. Jenkins, and P.F. Onyon, "*The Kinetics of Vinyl Polymerization by Radical Mechanisms*", Butterworths Sci. Publ., London 1961, p. 75.
[13] N. Tsubokawa, T. Kimoto, and K. Koyama, *Colloid Polym. Sci.* **1993**, *271*, 940.
[14] T.P. Almazova, E.P. Morozova, and V.I. Eliseeva, in: "*Composite Polymeric Materials*", Issue 24, Naukova Dumka, Kiev 1985, p. 61.

Use of Aminosilane Coupling Agents in Cementitious Materials

Franc Švegl, Jerneja Šuput-Strupi*

Slovenian National Building and Civil Engineering Institute, Dimičeva 12, SI-1000 Ljubljana, Slovenia
Fax: (+386) 1 2804484; E-mail: franc.svegl@zag.si

Summary: The work presented in this paper has been focused on the evaluation of the influence of the addition of aminosilanes like γ-aminopropyl-triethoxysilane (3-AMPS) on physical and mechanical properties of fresh and hardened polymer modified cement mortars (PMCMs), which contained also mineral fillers like fumed silica and blast furnace slag. The role of the aminosilane in cementitious system was to produce surface modification and covalent bonding among different inorganic phases of inorganic matrix (cement, aggregate, fumed silica) and strong chemical interaction between inorganic matrix and organic (polymer) phase of the mortar. The strong chemical bonding of inorganic matrix and polymer phase transformed composite mortar into cementitious material with more like monolithic structure. Mortars with the addition of aminosilanes showed improved physical-mechanical properties like bending strength, adhesion to concrete, and lower dry-shrinkage.

Keywords: aminosilanes; 3-AMPS; coupling agents; fumed silica; polymer mortars

Introduction

The repair of surface damaged concrete is usually accomplished with composite polymer modified cement mortars (PMCM's) consisting of series of constituents, which proportions determine mortars final properties. Nowadays is very popular to modify cement mortars with powdered acrylic or vinyl acetate based polymers and different types of fibers for reinforcement.[1,2] To achieve good performance of PMCM's a special attention should be devoted to ensure homogenous distribution of polymer powder and fibers in freshly prepared mortars. The addition of mineral fillers with pozzolanic activity, as fumed silica, blast furnace slag, and fly ash has significant effects on the physical and mechanical properties of cementitious materials. Fumed silica considerably improves compactness of cement based materials, which results in improvement of various properties like higher compressive strength, higher abrasion resistance, and lower

 DOI: 10.1002/masy.200550316

permeability of water and chemically aggressive substances. However, on the other hand, the negative influence of the silica fume is shown in increased drying shrinkage and lower workability of the fresh mortar or concrete.[3] An additional problem in designing of concrete and mortar mixtures containing mineral fillers poses rather low hydraulic activity of fumed silica and fly ash, which is much lower than that of Portland cement minerals.[4,5] To overcome these obstacles different organic admixtures are added to the fresh mortars, which predominately act as surface modifiers of cement and sand particles.[4] It has been shown that treatment of silica fume with mixture of Z-6020 and Z-6040 Dow Corning (Midland, MI) silane coupling agents improves physical and mechanical properties of fumed silica cement.[6] The basic idea of using silanes in cement based systems was taken from polymer composites where they serve as surface modifiers and coupling agents between polymer and mineral phase.[7,8] The use of aqueous amino silanes as an admixture to cement paste or cement mortar showed that they can completely substitute classical water-reducing agents with the additional improvement of workability, tensile and compressive strength, as well as some other physical and mechanical properties.[6,9-11] The combination of epoxy and amino silane coupling agents was successfully applied as an admixture to improve the degree of dispersion of carbon and steel fibers in fiber reinforced cement based composites.[1,2] The ability of aminosilanes to make covalent bridging between inorganic phases (e.g. cement, stone, fumed silica, slag) and strong chemical interaction between inorganic and organic phases (e.g. cement-polymer, fumed silica-polymer) to bind them together and form monolithic like structure makes them promising new type admixtures for cementitius materials. Concrete and mortars with the addition of aminosilanes are expected to show improved physical and mechanical properties such as lower water/cement ratio, higher flexural strength, better adhesion, lower shrinkage, better thermal conductivity, and better resistance to different sources of corrosion.

The main aim of this work was to evaluate the influence of the addition of water soluble γ-aminopropyl-triethoxysilane (3-AMPS) on physical and mechanical properties of fresh and hardened polymer modified cement mortars, which contained powder form ethylene-vinyl acetate (EVA) type polymer and mineral fillers like fumed silica and blast furnace slag. The laboratory testing of macroscopic physical and mechanical properties was accomplished according to the standard procedures for testing of mortars and cement (EN 1015, EN 1766, EN 196-1, and EN 196-3).

Experimental

Materials - The dry polymer modified mortar mixtures with different composition were prepared from Portland cement (CEM II/A-S 42,5 R, Cementarna Anhovo), fine quartz type aggregate, poly (ethylene-vinyl) acetate polymer (EVA), appropriate amount of fumed silica, and polypropylene fibers. The physical and mechanical properties, and chemical composition of Portland cement, CEM II/A-S 42,5 R, complied with the demands of standard EN 197-1. The fumed silica (FS) used was a dry uncompacted powder, which was applied as a partial replacement of cement by weight. The minimum content of SiO_2 in the fumed silica was 94 wt.%, the rest were iron and aluminum oxides. The polymer used for the modification of cement mortar was a milky-white powder of EVA having specific gravity of 1,150 g/cm^3 (20°C). The aminosilane coupling agent used as an admixture to PMCMs (up to 1 wt% to mass of cement) was liquid γ-aminopropyl-triethoxysialne (3-AMPS, Aldrich). The aminosilane was hydrolyzed prior the addition to cement based system at room temperature (at about 25°C) in highly alkaline environment and then polycondensed at elevated temperatures at 80°C for 4 hours to obtain colloidal solutions of pre-defined hybrid organic-inorganic particles or nanobuilding blocs (NBBs). Some of the mortars were reinforced with polypropylene fibers (PP) with the length of around 6 mm. Tensile strength of fibers was ranging between 540 to 750 MPa, their modulus of elasticity was around 3,400 MPa, the melting point was at 160°C and specific gravity was 0,91 g/cm^{-3}. A commercial melaminformaldehyde based superplasticizer (WRA) (ZETA PLUS, TKK Srpenica) powder was added to the mortar mixtures containing fumed silica to obtain a nominal slump or workability of around 160 mm.

Preparation of mortar - The fresh mortar mixtures with different formulation (Table 1) were prepared from dry premixed components in a stationary turbine mixer (Toni Technik/ToniMIX). The volume of the cementitious materials was held constant regardless of the aggregates and admixtures used. The water-cement ratios (w/c) ranged from 0,56 to 0,42. The mortars were mixed for 10 min. The sequence of batching and times of mixing were as follows: (i) mixing of dry mortar mixture (cement, coarse/fine aggregate, powder form admixtures, PP fibers and silica fume) for 2 min; (ii) adding water and mixing for 1 to 2 min; (iii) determining unit weight, slump, and air content; (iv) adding balance of water, and mixing for 2 min. The mass of the aggregate was determined in dry base, and the aggregate was then soaked in water for 24 h prior to mixing.

Preparation and curing of test samples for laboratory testing - Mortar prisms with dimensions 40 x 40 x 160 mm were molded for determination of compressive and flexural strength, and for determination of linear deformations. All the specimens were de-molded after curing in the chamber with relative humidity over 80% at 20 ±3 °C for 24 h. After de-molding, the prisms were cured under water at 20 ±3 °C for 3, 7, 28, and 90 days and for additional 7 days in air with 50 wt.% relative humidity prior to strength tests.

Testing of mechanical properties - The laboratory testing of compressive and flexural strength was performed in the Toni Technik/Toni Norm press. Measurements were accomplished according to the procedures described in standard EN 12190 and EN 196-1. The bonding strength of mortars with concrete plates with known characteristics was measured in a three point pull-off machine Josef Freundl Typ F15D EASY M.[11] Mortar samples were applied by hand in thickness up to 20 mm on standard reference concrete specimens prepared according to the procedure described in standard EN 1766. The method of test is by direct dolly pull-off using a circular dolly (diameter 50 mm) bonded to the surface of the applied mortar. The testing procedure is thoroughly described in standard EN 1542.

Specific heat - A Setaram Labsys DSC 131 Differential Scanning Calorimeter was used for measuring the specific heat of cement mortars containing polymer, silica fume and coupling agent. Sapphire was used as a reference material. The samples were taken from cured mortar prisms by cutting out small cylinders with diameter and height of around 3 mm. The sample weight was around 45 mg. The measurements were performed in the range from 25°C to 700°C at heating rate 10°C/min in Ar atmosphere (flow rate 10 ml/min).

Microstructural examination - Scanning electron micrographs of mortars were obtained on Jeol JSM-5500 LV scanning electron microscope at high or low vacuum. The cured prism samples of examined mortars were carefully fractured approximately at the middle of their length, applying minimum pressure to reduce the number of artificially generated cracks. The small cubic samples with dimensions 1 x 1 x 1 cm were prepared by cutting using an electric saw. Upon that some of the specimens were placed in plastic molds and vacuum saturated with low viscosity epoxy resin for 4 hours and subsequently placed in an oven set at 70°C to cure the epoxy. After the epoxy was fully cured, one face of the specimen was exposed by using a precision saw operated at low speed. The exposed face was ground and polished using cloth impregnated with diamond of successively finer sizes down to 0,25 μm. The exposed sample surface was coated with gold for examination under a scanning electron microscope equipped with an energy depressive X-ray analyzer.

Table 1. Mix proportions of dry powder mortars.

Mortar mixture	C/A*	EVA, wt. %**	Fumed silica, wt. %**	PP fibers, wt. %**	3-AMPS, wt. %**	WRA, wt. %**
CM	1/3	-	-	-	-	-
PM	1/3	5	-	-	-	-
PMA	1/3	5	-	-	0,6	-
PMSF	1/3	5	10	-	-	1,5
PMSFA	1/3	5	10	-	0,6	-
PMSFPPF	1/3	5	10	0,5	-	1,5
PMSFAPPF	1/3	5	10	0,5	0,6	-

* - C/A is a ratio between cement and aggregate.

** - All materials are expressed in wt. % with respect to the mass of cement.

CM - reference cement mortar containing only Portland cement CEM II/A-S 42,5 R and fine type quartz aggregate without any additives.

PM - polymer modified cement mortar.

PMA - polymer modified cement mortar containing 3-AMPS coupling agent.

PMSF - polymer modified cement mortar containing fumed silica filler and water reducing agent (WRA).

PMSFA - polymer modified cement mortar containing fumed silica and3-AMPS coupling agent.

PMSFPPF - polymer modified cement mortars containing fumed silica, PP fibers and water reducing agent (WRA).

PMSFAPPF - polymer modified cement mortars containing fumed silica, 3-AMPS coupling agent and PP fibers.

Results and Discussion

Workability – The workability of mortars was determined at constant water/cement ratio (w/c) of around 0,55 by the slump test described in the standard EN 1015-3. The results in Table 2 show that the addition EVA polymer increases the workability of mortar according to expectations and in agreement with the data from the literature.[12] The high demand for water of the fumed silica and resulting decrease of mortar workability was compensated by the addition of water reducing agent (WRA). The addition of 3-AMPS increases the workability of polymer modified mortar and even more of polymer modified mortar containing mineral admixtures such as fumed silica and blast furnace slag. The workability of mortars containing aminosilane (up to 0,6 wt.% by weight of cement) was better than the mix with as-received silica fume and water reducing agent (1,5 wt.% by the amount of cement). With coupling agent and water-reducing agent in the amount of 0,1 wt.% by the amount of cement, the workability was better as with mortars containing fumed silica and only water-reducing agent in the amount of 1,5 % by weight of cement. The results showed that it is possible to completely substitute superplasticizer (WRA) by an aminosilane additive. The high workability of mortars containing aminosilane can be explained by its hydrophilic nature due to the presence of amino and hydroxide groups.

Table 2. Properties of fresh mortars.

Mortar mixture	Slump, mm	Unit weight, kg dm^{-3}	Air void content, %
CM	132	2,17	5,1
PM	154	2,06	6,8
PMA	182	2,14	5,1
PMSF	155	2,01	9,1
PMSFA	188	2,19	4,2
PMSFPPF	175	2,04	6,9
PMSFAPPF	183	2,18	4,3

Air void content and density – The air void content and density of mortars were determined by the procedures described in standard EN 1015-6,7. The most pronounced increase in air-void content was observed by adding fumed silica to the mortar (Table 2). The accompanying effect is a decrease in density of fresh mortars. The addition of aminosilane decreased the air void content of mortars containing fumed silica for more than 100 %, which resulted in considerable increase of density of the fresh mortar mixtures. The explanation for this effect is in surface modification of cement and sand particles by aminosilane molecules, which enables strong chemical interaction between mineral-mineral and mineral-polymer phases resulting in more compact structure with less air entrapped in the solid matrix.

Mechanical properties – As shown in Figure 1 and 2 for mortars after 3, 7, 28, and 90 days of curing under water and 7 days in air, the compressive strength and bending strength are increased in early stage of hydration of the cement. A considerable decrease in compressive strength of polymer modified mortar, PM, was observed after longer time period (90 days). This shows the degradation of EVA polymer in highly alkaline environment of cement. The addition of fumed silica (10 wt.% by weight of cement) increased the compressive strength as expected.[13] The presence of 3-AMPS in polymer modified mortar improved the stability of EVA polymer and its performance with time. In general, the compressive strength was not much affected by the addition of aminosilane to the mortars (Figure 1). However, the bending strength of all mortars was considerably increased when 3-AMPS was present. The bending strength after 28 days for mortars containing 3-AMPS was increased for more than 60 % in comparison to reference cement mortar (CM). The addition of 3-AMPS to the mortars containing PP fibers (up to 0,5 wt. % to the mass of cement) improved the homogeneity in dispersing of fibers through the whole volume of the mortar, which resulted in additional improvement of bending strength. It is assumed that the hydrocarbon part in H_2N-CH_2-CH_2-CH_2- chain of 3-AMPS influenced the wettability of the PP fibers and that the chemical affinity of cementitious system to polypropylene was improved.

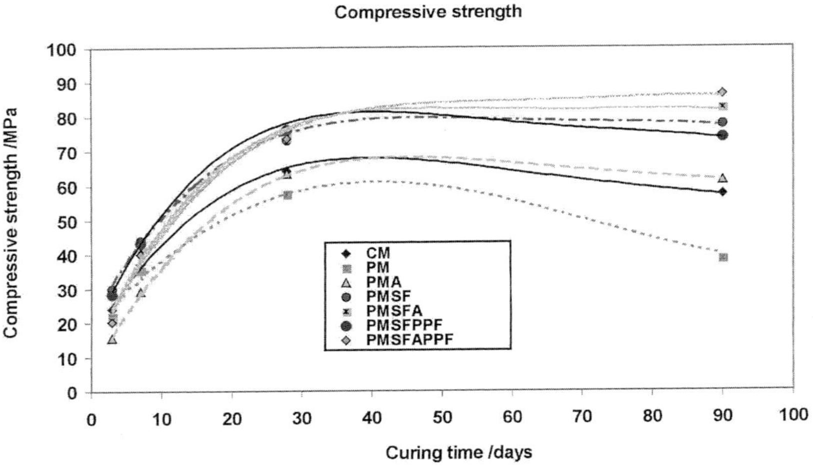

Figure 1. Compressive strength of mortars after 3, 7, 28, and 90 days curing under water and additional 7 days in air.

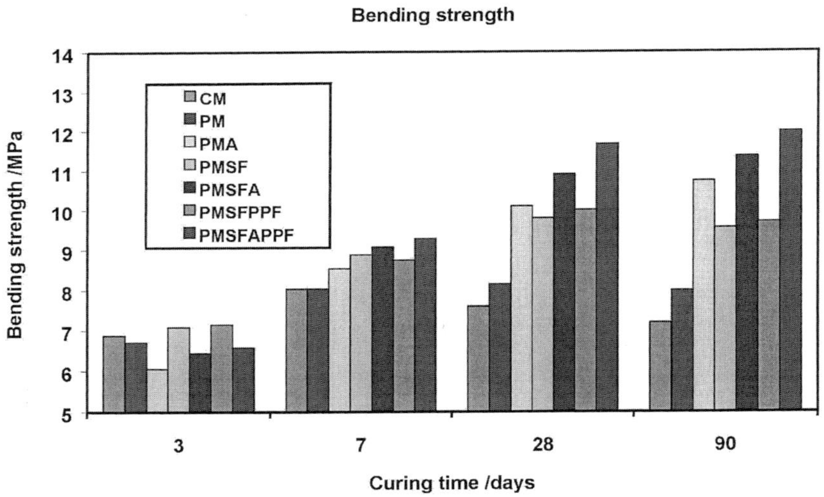

Figure 2. Bending strength of mortars after 3, 7, 28, and 90 days curing under water and additional 7 days in air.

Drying shrinkage – The results of the measurements of linear deformations of mortar prisms after drying in air for 28 days showed that the addition of EVA polymer and fumed silica reduced the drying shrinkage (Table 3). The addition of aminosilane has an additional beneficial effect on drying shrinkage by reducing linear deformations of prisms for more than 40 %. The formation of covalent chemical bonds of 3-AMPS at the interfaces of inorganic phases compensates stresses occurring due to different shrinkage characteristics of the mortars components.

Specific heat – The results of specific heat measurements by DSC of cement pastes modified with the same amount of additives showed that the addition of polymer and fumed silica increased the specific heat of mortars depending on the amount of added modifier (Table 3). When aminosilane was added to the cement paste the formation of the network of covalent bonding between particles of cement greatly contributes to the absorption of energy due to the increased amount of phonon vibrations. This resulted in considerable increase of specific heat, which was above 0,6 wt. % of added 3-AMPS to the mass of cement independent on the amount of added aminosilane.

Table 3. Linear deformations of mortars after 28 days drying in air and specific heat of cement pastes with the addition of polymer, fumed silica, and aminosilane.

Mortar mixture	Linear deformations, mm/m	Specific heat, J/g K
CM	-0,95	0,698
PM	-0,84	0,746
PMA	-0,67	0,883
PMSF	-0,90	0,921
PMSFA	-0,53	1,121
PMSFPPF	-0,74	0,933
PMSFAPPF	-0,45	1,119

Adhesion to concrete – The results of adhesion testing with pull-off method showed that the addition of fumed silica considerably improved the bonding properties of mortars to concrete substrate. This positive effect can be attributed to the fact that fine particles of fumed silica easily fill the pores of concrete even few millimeters into the bulk, which increased the contact surface area of mortar with concrete. The addition of the aminosilane coupling agent dramatically improved the adhesion of the mortars to concrete (Figure 3). This showed that 3-AMPS enabled even more efficient contact and stronger bonding between the particles of mortar and concrete substrate. The visual assessment of the type of failure categorized according to the standard EN 1542 showed that with all specimens containing silica fume and aminosilane cohesion failure in the concrete substrate (type A) occurred.

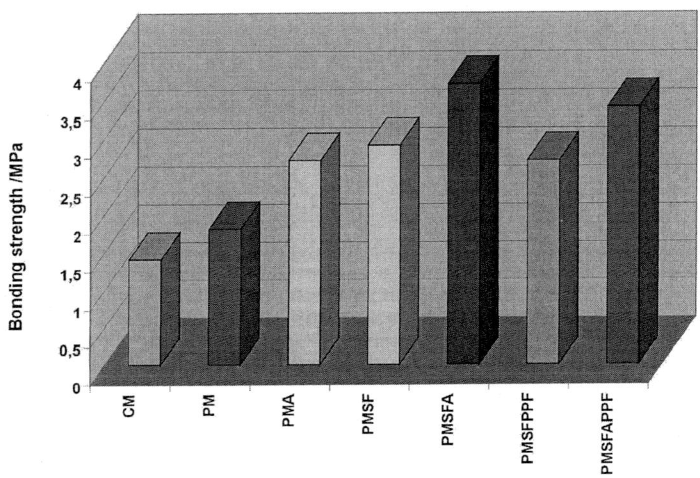

Figure 3. Pull-off results of mortars deposited on concrete panels obtained after 28 days.

Morphology - SEM micrograph of polymer mortar containing fumed silica (PMSF) (Figure 4a) showed rather compact structure with small cracks appearing between aggregate particles and cement matrix. From the SEM micrograph of the mortar containing fumed silica and aminosilane (PMSFA) (Figure 4b) it can be seen that almost all the voids and cracks disappeared. The borders between stone particles and cement phase are less distinct as with mortars without the addition of aminosilane. This indicates that coupling agent improved the adhesion between neighboring particles and bound them together into three-dimensional, dense, monolithic like structure.

(a) (b)

Figure 4. (a) SEM micrograph of the polymer mortar containing fumed silica (PMSF). (b) SEM micrograph of the polymer mortar containing fumed silica and aminosilane (PMSFA).

Conclusions

The incorporation of mineral fillers with pozzolanic activity as fumed silica had great positive impact on physical and mechanical properties of polymer modified cement mortars. An additional beneficial effect was obtained by introduction of aminosilane additives based on 3-AMPS. The introduction of aminosilanes as surface modifiers and coupling agent was a new approach in modifying of cementitious systems. The major impact of the addition of 3-AMPS was observed in improving bending strength, drying shrinkage and adhesion properties of mortars to concrete.

Acknowledgments

The authors gratefully acknowledge the financial support from Ministry of Education, Science and Sport, Slovenia. This work was accomplished in the scope of the project L2-4520.

[1] Y. Xu, D. D. L. Chung, *Cem. Cocn. Res.*, **1999**, 29 (5), p.773.
[2] J. Cao, D. D. L. Chung, *Cem. Cocn. Res.*, **2001**, 31, p.309.
[3] J. M. Gao, C. X. Qian, B. Wang, K. Morino, *Cem. Conc. Res.*, **2002**, 32 (1), p.41.
[4] F. Massazzana, M. Moranville-Regourd, P. Fidyestøl, R. Lewis, in *"Leas's Chemistry of Cement and Concrete"*, Ed. by P. C. Hewlett, 4th Edition, Arnold, 1998**,** Chapter 10-12, p.471.
[5] H. F. W. Taylor, *"Cement Chemistry";* Academic Press, London, 1990, Chapter 7, p.199.
[6] Y. Xu, D.D.L. Chung, *Cem. Conc. Res.*, **2000**, 30, p.1305.
[7] E. P. Plueddemann, *"Silane Coupling Agents"*, Plenum Press, New York, 1982, Chapter 3, p.49.
[8] A. V. Pocius, *"Adhesion and Adhesives Technology"*, Carl Hanser Verlag, Munich, 2002, Chapter 6, p.153.
[9] F. Švegl, J. Šuput-Strupi, *Proceeedings of the Worshop in Chemistry*, Maribor, Slovenia, 20-21 September **2001**, p.630.
[10] F. Švegl, J. Strupi-Šuput, S. Kadunc-Kodra, J. Grdadolnik, *Proceedings of the Workshop in Chemistry*, Maribor, Slovenia, 26. and 27. September, **2002**, p. 845.
[11] F. Švegl, J. Strupi-Šupput, *Proceedings EUROFILLERS 2003,* Alicante, Spain, 8-11 September, **2003**, p. 203.
[12] M. U. K. Afridi, Z. U. Chaudhary, Y. Ohama, K. Demura, M. Z. Iqbal, *Cem. Conc. Res.*, **1994**, 24, p.1199.
[13] Z. Bayasi, J. Zhou, *ACI. Mater. J.*, **1993**, 90, p.349.

Synthetic Hydromagnesite as Flame Retardant. A Study of the Stearic Coating Process

Laia Haurie,[1] *Ana Inés Fernández,**[1] *José Ignacio Velasco,*[2] *Josep Maria Chimenos,*[1] *Josep Ramón Ticó-Grau,*[3] *Ferran Espiell*[1]

[1] Unitat de Ciència de Materials i Enginyeria Metal·lúrgica, Universitat de Barcelona, Martí i Franquès, 1, 08028 Barcelona, Spain
E-mail: frenna@material.qui.ub.es
[2] Centre Català del Plàstic, Universitat Politècnica de Catalunya, C/ Colom, 114, 08222 Terrassa, Barcelona, Spain
[3] Tecnologia Farmacèutica, Facultat de Farmàcia. Universitat Barcelona, Avda. Joan XXIII s/n, 08028 Barcelona, Spain

Summary: A synthetic hydromagnesite obtained from an industrial by-product rich in magnesium oxide was employed and evaluated as a non-halogenated flame retardant for poly(ethylene-co-vinyl acetate).
The filler was characterized with different techniques (such as specific surface area, TGA, particle morphology and size measurements, WAXS). Significant differences were found between the synthetic hydromagnesite and the natural one.
Synthetic hydromagnesite was coated with stearic acid and the effectiveness of the coating process was studied by the dye adsorption method and sedimentation volume measurements. The amount of coating agent ranged from 1 to 4.5%. This factor was found to have a significant effect on the thermal decomposition behaviour of the filler.
A poly(ethylene-co-vinyl acetate) (27% of VA) was filled with the coated synthetic grades of hydromagnesite as well as with two commercial flame retardants and different physicochemical properties were evaluated, including their flame retardant effect.

Keywords: coatings; fillers; flame retardance; thermogravimetric analysis (TGA); thermoplastics

Introduction

Hydromagnesite, a basic magnesium carbonate, is referred to in literature[1] as a flame retardant filler for polymers in cable applications. The synthetic hydromagnesite used in this study is a basic magnesium carbonate, obtained by Magnesitas Navarras S.A from an industrial by-product rich in magnesium oxide[2], with a chemical composition corresponding with the formula $5MgO_4CO_25H_2O$. In the temperature range of 200-550°C it undergoes an endothermic decomposition with an associated heat of 800 J/g

 DOI: 10.1002/masy.200550317

and a final weight loss of 54%. The steps of this decomposition reaction are shown below; as can be seen there is an initial release of water in the form of steam followed by a final step where carbon dioxide is evolved and magnesium oxide is formed [3].

$$4MgCO_3 \cdot Mg(OH)_2 4H_2O \rightarrow 4MgCO_3 \cdot Mg(OH)_2 + 4H_2O$$

$$4MgCO_3 \cdot Mg(OH) \rightarrow 4MgCO_3 \cdot MgO + H_2O$$

$$4MgCO_3 \cdot MgO \rightarrow 5MgO + 4CO_2$$

This endothermic breakdown, as well as the secondary effects like dilution and cooling of the gas phase by water vapor and carbon dioxide released and formation of ceramic substrates, make hydromagnesite a potential flame retardant in the polymer field, acting with a mechanism similar to those of metal hydroxides like aluminium or magnesium hydroxide [4,5].

Experimental

The process of obtaining hydromagnesite involves the formation of a $Mg(HCO_3)_2$ solution by carbonation of a MgO-containing residue slurry and further precipitation of hydromagnesite by addition of pure magnesium oxide as the precipitating agent [6].

In order to achieve a good performance as flame retardant, a high loading of hydromagnesite in the plastic system is usually required. Surface modification with fatty acids is used to improve polymer processing and final properties of polymer compounds, as well as reducing the agglomeration tendency of the filler [7,8].

The dry blending coating process was developed in a double jacket high-speed homogenizer initially heated at 40-45°C. The temperature was increased up to 60°C and afterwards the system was slowly cooled to room temperature. The stearic acid used was of pharmaceutical grade provided by Roig Farma. The amount of coating agent was ranged from 1 to 4.5% to evaluate the influence of this factor on the thermal decomposition behaviour of the hydromagnesite and the compounds prepared thereof.

The poly(ethylene-co-vinyl acetate), EVA, (Escorene Ultra UL00728, 27% of VA) and the different coated grades of hydromagnesite were mixed in a two roll mill at 75-80°C. The obtained compounds were compression moulded to produce 1 mm thick sheets. In order to establish a comparison with commercial inorganic flame retardants, compression moulded sheets were prepared with Magnifin H5 KV, a magnesium hydroxide grade with an organic surface treatment, and Ultracarb C5-10, a natural obtained hydromagnesite and huntite coated with 1% of stearate.

Specific density, specific surface and particle size distribution and morphology were measured in order to characterize the synthetic coated and uncoated hydromagnesite. Specific density was measured with an helium picnometer and specific surface area by the single point BET method. Particle size distribution was determined by light scattering with a Beckman Coulter LS 13 320 apparatus. Particle morphology could be observed by scanning electron microscopy (SEM).

The effectiveness of coating was evaluated by the dye adsorption method using red acid Ponceau 3R, known to adsorb onto $Mg(OH)_2$[9]. An excess of dye was added to the sample and the unbound mass was determined by visible spectroscopy using a Shimadzu UV-2101 PC spectrometer working in the range of 190-800nm.

Sedimented volume measurements of the different hydromagnesite grades in hexane were also carried out to evaluate differences in the behaviour between treated samples. The experimental procedure is accurately described in literature [9].

Thermogravimetric analyses (10°C/min, O_2 atm) were performed in a Setaram TG DTA92 thermobalance. Wide-angle x-ray scattering (WAXS) was employed to get information about the crystalline phases (magnesium oxide) formed during the hydromagnesite thermal decomposition.

Flame retardant filled EVA samples were characterized by values of powdered filler concentration, density, Vicat softening point, hardness, tensile properties and SEM micrographies of the surface fracture were obtained. Furthermore, values of oxygen index and dropping time after flame application were obtained to evaluate the flame retardant behaviour of these EVA compounds. The average dropping time (DT) was measured as the time that a specimen of dimensions 120 mm x 10 mm x 1 mm takes to drop after applying a normalized flame with bottom ignition for 10 seconds.

Results and Discussion

Hydromagnesite samples are coded as follows: uncoated hydromagnesite (UH); surface coated samples of hydromagnesite with different concentration of stearic acid ranged from 1 to 4.5 wt% (H1-H4.5).

The density mean value of UH was 2.43 g/cm^3. In surface-treated filler samples this value was found to be reduced to 2.27 g/cm^3 independently of the percentage of coating applied. The specific surface area was not modified by the coating, all the samples having a value near to 24 m^2/g.

The coating effectiveness parameter was defined as the mass of adsorbed red acid relative to the initial dye mass (χ). Stearic acid presented no dye adsorption, while hydromagnesite showed affinity to adsorb acid dyes like Ponceau 3R. The amount of adsorbed dye decreased significantly beyond 2% of stearic acid, which could be due to a variation of the free hydromagnesite specific surface. Values of sedimented volume decreased with coating agent added up to 3% of stearic acid, probably due to a polarity reduction, and then remained steady.

Table 1. Hydromagnesite grades values of density, sedimented volume and coating effectiveness parameter.

Samples	Density (g/cm^3)	V_{sed} (ml)	X
UH	2.43	53.0	0.34
H1	2.27	44.0	0.33
H2	2.27	27.0	0.32
H3	2.26	24.0	0.22
H4	2.28	23.5	0.25
H4.5	2.27	24.0	0.26

Thermogravimetric analyses (TGA) were performed to study the thermal behaviour of the samples. As mentioned before, thermal decomposition of uncoated hydromagnesite follows three steps within the temperature ranges of 210-250°C, 370-420°C and 480-530°C. The effect of increasing the amount of stearic acid has been related with changes in the thermal decomposition kinetics of the hydromagnesite, as is depicted in the TGA shown in Figure 1.

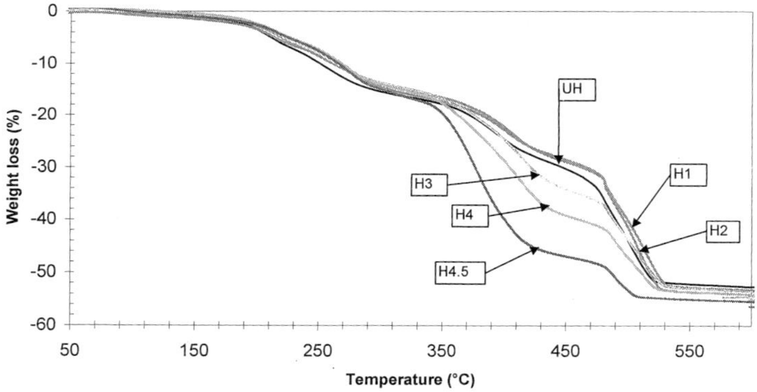

Figure 1. TGA plots of hydromagnesite grades.

Substantial differences were found in the temperature range from 320 to 480 °C corresponding to decomposition steps that result in H_2O and CO_2 liberation, and oxide formation. A remarkable increase in the decomposition rate was observed at the beginning of that interval as the coating level was increased in the filler. The filler uncoated and coated with the lower levels of stearic acid (< 2 wt %) did not follow this trend, having similar behaviour to the unmodified filler.

To analyse the differences in the oxide formation, UH and H4.5 samples were heated at 400°C for one hour, afterwards WAXS patterns were recorded (Figure 2). The MgO presence was observed in both samples, but the H4.5 sample clearly showed a more intense and defined MgO signal. The shape of the diffraction peaks suggest higher crystallinity in MgO formed from the H4.5 sample, which would be in good agreement with a higher oxide formation rate.

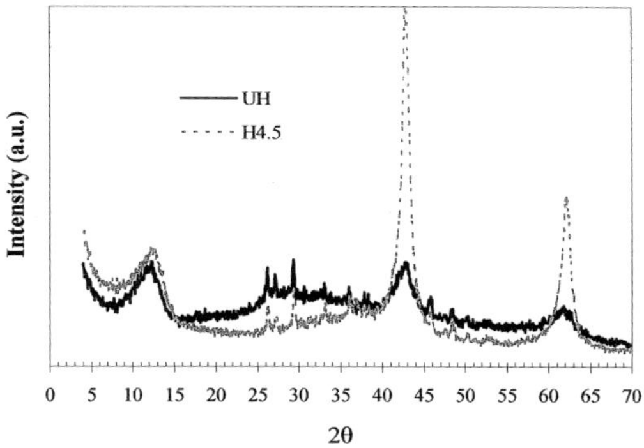

Figure 2. X-Ray patterns of samples after thermal treatment at 400°C.

Aspects like particle size and morphology usually require to be controlled to guarantee optimal final properties of the polymer compounds.

Numerical particle size distribution showed no significant differences among uncoated hydromagnesite and coated samples, giving a mean particle size (d_{50}) of 0.1μm for UH and 0.09μm for the coated grades. However, the volumetric distribution exhibited higher particle size values (d_{50} = 13.3 μm for coated particles; d_{50} = 16 μm for UH). This discrepancy could probably be due to the presence of agglomerates. Also, the volumetric distribution was found to be broader than the numerical one, with the d_{90} values 57μm and 38μm for uncoated and coated filler respectively. This improvement may be attributed to a decrease in the level of agglomerates because of the coating process. In Figure 3 a SEM micrograph of H1 sample is shown. The morphology of these particles forming fine platelets can be observed.

Figure 3. General aspect and magnification of H1particles by SEM.

A mixture of hydromagnesite with huntite, named Ultracarb®, was also studied for comparison. It presented a bimodal size distribution, Figure 4. The first curve reported values smaller than one micron and the second had a mean value near 3µm. As can be seen in Figure 4, morphological aspect of the Ultracarb® particles, forming thin platelets, was similar to the particle shape of synthetic hydromagnesite.

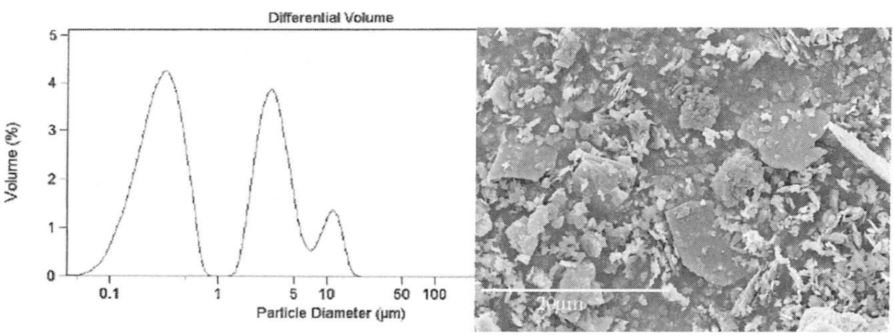

Figure 4. Ultracarb volumetric particle size distribution and SEM image.

The presence of aggregates could result in a poor homogeneity in the filled polymer, affecting its mechanical properties. In order to improve this aspect the optimal milling time for the hydromagnesite was studied. In Figure 5 volumetric particle size distributions of hydromagnesite samples milled for 0.5 and 5 minutes are compared.

The improvement obtained with a 5 minutes grinding step is also confirmed with the SEM picture presented in Figure 5. Higher milling times do not present significant reduction in the particle size.

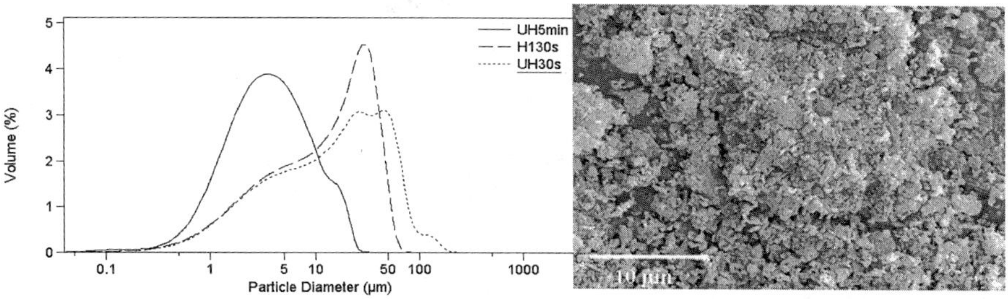

Figure 5. Variation of hydromagnesite particle size and volumetric distribution with milling time.

The nomenclature employed for the filled polymers was as follows: EVA filled with uncoated hydromagnesite (EUH); EVA filled with hydromagnesite treated with stearic acid from 1 to 4.5wt% (EH1-EH4.5); EVA filled with Ultracarb C5-10 (EU) and EVA filled with Magnifin H5 KV (EMn).

Table 2 summarizes some relevant characteristics of the compounds. As expected, the density value was in concordance with the filler concentration determined by ashing. The Vicat temperature was almost independent of the filler loading, although a slight decrease seemed to be observed due to the high surface coating levels.

The highest DT value was found for EH2 sample. Despite DT values show considerable error limits (between 5-15%), this good fire response was confirmed by the limiting oxygen index results. The oxygen index found for sample EH2 was 32.9. From this point a progressive decrease could be observed as coating level was increased.

There were many factors affecting the fire behaviour of these samples, the most important ones being the filler loading and the percentage of added coating agent. This last factor being responsible for the filler decomposition changes as mentioned before.

Table 2. Concentration, density, Vicat softening temperature (VST), dropping time (DT) values and oxygen index (OI) measured for hydromagnesite filled EVA compounds.

Sample	Filler (%)	ρ (g/cm³)	VST (ºC)	DT (s)	OI (%)
EUH	50	1.29	54.3	14	24.5
EH1	55	1.37	56.9	21	31.9
EH2	57	1.43	53.8	38	32.9
EH3	56	1.40	53.4	23	29.8
EH4	57	1.45	53.9	19	29.2
EH4.5	56	1.37	50.4	17	27.6

Standard tensile specimens were machined from the moulded plaques and tests were performed at room temperature and at a cross-head speed of 10 mm/min. Although tensile properties (Table 3) were highly dependent on the filler loading substantial differences related to the coating concentration were noticed. Secant elasticity modulus at 1% of strain ($E_{1\%}$) as well as hardness values increase with filler loading up to 2% of stearic acid with a progressive reduction of these values for samples EH3 and EH4.5 that could indicate a plasticizer effect of the stearic acid on EVA compounds. This apparent plasticizing effect could also justify the higher elongation at break of samples with an excess of coating agent. A high value of elongation at break (70%) resulted for EUH sample. The cause of that would be the low filler content (50%) in this compound. Tensile strength (σ) increased up to 2-3% of stearic acid and showed a reduction in samples with an excess of this coating agent. Higher adhesion between filler and matrix would explain the increase in tensile strength. A SEM study of the tensile fracture surfaces was carried out. In Figure 6 the interface adhesion induced by the stearic acid on the EVA-hydromagnesite compounds can be observed.

The comparative results obtained with the EVA compounds filled with the different inorganic flame retardants are summarized as follows. Synthetic hydromagnesite compounds had higher $E_{1\%}$ and σ, but less deformability than Ultracarb and Magnifin compounds (Table 4). Furthermore, as depicted in Figure 7, higher deformability of the polymer matrix was observed for EMn and EU samples. Differences observed in $E_{1\%}$ may be attributed to the differences among the fillers such as particle size distributions or coating treatment. The low value of $E_{1\%}$ in EMn could be related with the surface coating of this filler with a silicone derivate. On the other hand, the good adhesion of coated hydromagnesite-EVA compounds observed by SEM would be responsible of the higher value of $E_{1\%}$. Dropping time and oxygen index values were strongly affected by the filler content.

Table 3. Mechanical properties of filled EVA.

Samples	$E_{1\%}$ (MPa)[a]	σ (MPa)[b]	ε (%)[c]	Hardness (Shore D)
EUH	107	6.1	70	40
EH1	180	7.6	27	44
EH2	222	8.7	46	47
EH3	181	8.4	54	46
EH4	162	7.6	52	44
EH4.5	165	7.7	80	42

[a] Secant elasticity modulus at 1% of strain. [b] Tensile strength. [c] Elongation at break.

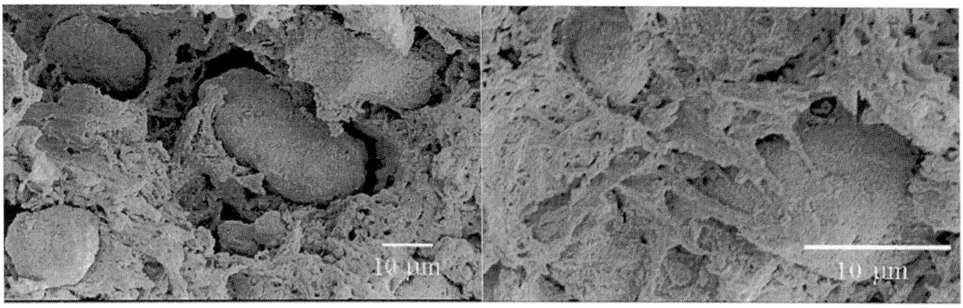

Figure 6. Fracture surfaces of EUH and EH1 showing different adhesion particle-matrix.

Table 4. Characterization of hydromagnesite, Ultracarb and Magnifin filled EVA compounds.

Samples	Filler (%)	ρ (g/cm^3)	VST (°C)	DT (s)	OI (%)	$E_{1\%}$ (Mpa)	σ (Mpa)	ε (%)
EH1	55	1.37	56.9	21	31.9	180	7.6	27
EU	63	1.51	57.4	45	34.7	122	6.5	149
EMn	60	1.46	57.4	19	39.2	75	6.7	204

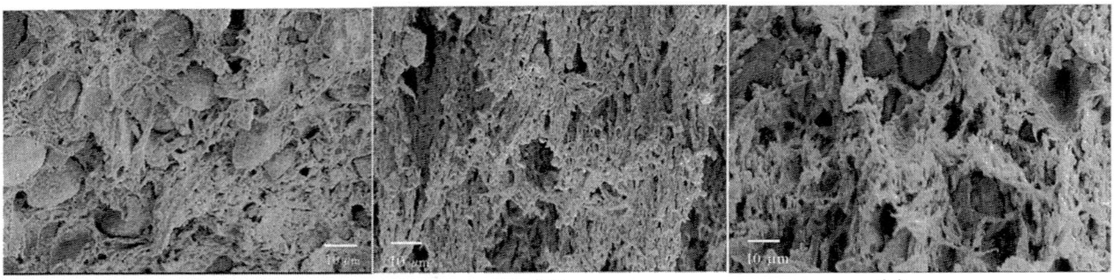

(a) (b) (c)

Figure 7. Fracture surface SEM images of EH1 (a), EMn (b) and EU (c).

Despite this, it could be considered that the fire performances of the synthetic hydromagnesite compounds were comparable with the commercial flame retardants studied. Reduction of filler particle size and therefore a narrow distribution could improve mechanical properties of synthetic hydromagnesite filled EVA compounds.

Conclusions

Synthetic hydromagnesite has proven to be suitable filler for flame retardant polymer formulations, comparable to commercial grades based on natural hydromagnesite as well as to other inorganic flame retardant fillers.

An optimal content of stearic acid as coating agent was found within the range 1-2%. Higher percentages of stearic acid seemed to induce adverse effects during the thermal decomposition of hydromagnesite and on the flame retardant behaviour of filled EVA compounds.

The homogeneity of synthetic hydromagnesite particle size distribution was remarkably improved by the addition of a controlled milling step in the production process.

Acknowledgments

The authors would like to thank Magnesitas Navarras S.A for supporting this work and supply the samples. The authors also thank Mr. T. Aliari and Mrs. M. Salinas for his assistance with experimental procedures and Mr. D. Binagui for his technical support.

[1] H.A. Mayer, B.K. Harmsworth, K.G. Cooper, WO 0226879, 2002
[2] A.I. Fernández, J.M. Chimenos, M.Segarra, M.A Fernández, F. Espiell, *Ind. Eng. Chem. Res.* 2000, 39, 3653
[3] G.S Kirschbaum, *Flame retardants 98*, Proceeedings of the Conference, January 1998, 151
[4] M. Rigolo, R.T. Woodhams, Polymer Engineering and Science, 1992, 32, 327
[5] G.J Simandl, J. Simandl, A. Debreceni, Geological Fieldwork 2000, 2001, 327
[6] F. Espiell, J.M. Chimenos, M.A Fernández, M.Segarra, A.I. Fernández, ES 2 141 677, 2000
[7] R.N. Rothon, Fillers Symposium, Loughborough, UK, 1996
[8] M. Gilbert, P. Petiraksakul, I. Mathieson, Materials Science and Technology, 2001, 17, 1472
[9] C.M. Liauw, R.N. Rothon, S.J. Hurst, G.C Lees, Composite Interfaces, 1998, 5, 503

Macromol. Symp. **2005**, *221*, 175-184

The Effect of Glass Fibre and a Phosphorus-Containing Flame Retardant on the Flammability of Recycled PET

Maria Lluïsa Maspoch,[1] Haritz E. Ferrando,[1] Diego Vega,[2] Antonio Gordillo,[1] José Ignacio Velasco,[1] Antonio B. Martínez[1]*

[1] Centre Català del Plàstic (CCP), Universitat Politècnica de Catalunya (UPC), Colom 114, 08222 Terrassa, Spain
Fax + 34 93 784 18 27; E-mail: maria.lluisa.maspoch@upc.es
[2] SONY España S.A. Barcelona Plant, Pol. Ind. Can Mitjans, 08232 Viladecavalls, Spain

Summary: The availability of scrap poly(ethylene terephthalate) (PET) from post-consumer bottles is increasing as the post-consumer collecting systems are becoming more usual in daily life. PET is well known as a high-performance engineering thermoplastic because of its good thermal stability, chemical resistance, and excellent mechanical properties. Many efforts have been carried out to use this material in housings of electronic applications. However, the flammability of PET is a shortcoming in some of these applications.
In this study, our attempt is to incorporate a non-halogenated flame retardant, in form of a phosphorus-containing compound, together with a commercial glass fibre grade to achieve UL94 test V-0 rating for PET. An investigation of thermal stability and flammability (HDT, UL94 V-test) and mechanical (tensile, flexural and impact tests) properties of glass fibre filled PET samples is reported as a function of fraction of flame retardant.
This work shows the influence of the filler content and the interfacial filler/matrix adhesion on the flame retardant and the mechanical properties.

Keywords: fibres; flame retardance; mechanical properties; phosphorus; recycling

Introduction

Recycling of polymeric materials is of fundamental importance because of environmental, economic and social factors, and in recent years has grown in a very impressive way. Concerning mechanical recycling of thermoplastics, studies are focused on the thermo-mechanical degradation processes appearing during recycling processes resulting in a poor secondary material, with regard to its properties [1].

DOI: 10.1002/masy.200550318

The availability of scrap poly(ethylene terephthalate) (PET) from post-consumer bottles is increasing as the post-consumer collecting systems are becoming more usual in daily life. PET is well known as a high-performance engineering thermoplastic because of its good thermal stability, chemical resistance and excellent mechanical properties. Since recycled PET is becoming more competitive in some sectors, replacing more common polymers, many efforts have been carried out to use this material in applications considered in this study (TV components) [2-3].

In this study, the flammability of PET is a shortcoming in some of these applications. Requirements about burning behaviour and the need of using an halogen-free flame retardant, for environmental and public health reasons, have led to the study of the addition of a phosphorus-based flame retardant in recycled PET [4-5].

The aims of this study are to: investigate the synergistic effect of a phosphorus-containing flame retardant and glass fibre on the flammability of PET; optimise additive contents of the mixtures to obtain UL94 V-0 rating; and study the influence of the additives on the mechanical properties.

Experimental

Materials

The recycled material of this study is a mixture of PET flakes of different colours, coming from post-consumer sorted plastic bottles (PET-B). A commercial PET, provided by DuPont, RYNITE R 530L NC 010, filled with 30% glass fibre (PET-A), has been used as a reference to investigate the role of glass fibre interfacial adhesion on the flame retardant behaviour.

The filler fraction consists of two types of glass fibre (GF), one type, class E, without a surface treatment, and a second type FV Chopped Strand 3540 from PPG Industries, provided by QUIMIDROGA, with a surface treatment adapted for PET (no details of the coupling agent are provided).

The flame retardant (FR) is a viscous liquid phosphorus compound FYROFLEX BDP Bisphenol A bis-(diphenylphosphate), provided by AKZO NOBEL, with a phosphorus-content of 9%.

Processing

PET materials were dried at 140°C for 4 hours prior to and during extrusion and injection moulding processing in a dehumidifying drier. Masterbatches of 35% weight GF with PET-B and masterbatches of 15%GF + 53%FR with PET-B were made by twin-screw extrusion (COLLIN ZK-25, L/D=36). The processing temperatures were 250-270°C depending on the compositions. Both masterbaches were physically blended in different proportions with PET-B flakes to prepare the materials of the study (table 1). All materials were injection moulded in a mould cooled at 20°C to obtain standard specimens (figure 4 of the ASTM D-647 standard). The specimens produced were dumbbell for tensile tests and prismatic (3.2*12.7*127 mm) for impact, HDT and flammability tests.

Characterisation

Flammability tests were performed according to UL94 norm at room temperature. A sample was exposed vertically to a Bunsen burner flame for 10 seconds. If the flame extinguished, another 10 seconds application was made. Total burning time of the two ignitions for five samples are shown in table 1. Charpy impact tests were carried out according to ASTM D-256 using a CEAST pendulum impact tester with 2 J capacity providing an impact speed of 2.9 m/s on specimens with a 1.25 mm notch. Determination of HDT temperature was performed according to ISO 75 in a CEAST 6510, each material 5 times. Tensile tests were performed according to ASTM D-638M-91a in a Galdabini 1890 Sun 2500 universal testing machine with a load cell of 25 kN and a video extensometer, at a crosshead speed of 10 mm/min at room temperature.

Results and Discussion

Due to confidential reasons, phosphorus and glass fibre contents are shown relative to values of PET-C5 phosphorus content. The selection of this material as a reference is because it shows the best flame retardance behaviour with the minimal phosphorus content.

Preliminary study of the upward-burning characteristics

Firstly, the study has investigated the flammability of different starting materials.

Flammability of PET-B is V-2 rating. It shows considerable dripping and ignites the surgical cotton. Previous studies concluded that addition of a phosphorus-containing flame retardant alone, even in high concentrations, did not lead to any improvement in flame retardation, relative to the original material (PET-B). Furthermore, addition of untreated glass fibres (PET-D) did not benefit flame retardation at all; the dripping behaviour persisted such that the material had the same UL94 classification as the unfilled matrix i.e., UL94-V2. A study of the flammability of PET-A showed that the structure of the filler phase was maintained after the flammability test and even after complete removal of the organic phase at 600°C. Thus, the treatment of the glass fibre with a coupling agent was considered as an important factor of the study, to reach a better flammability behaviour. It may be noted in the Scanning Electron Microscopy (SEM) micrographs that the interfacial adhesion between the PET matrix and the glass fibre phase does not exist in material PET-D (figure 1), while it is effective in material PET-A (figure 2).

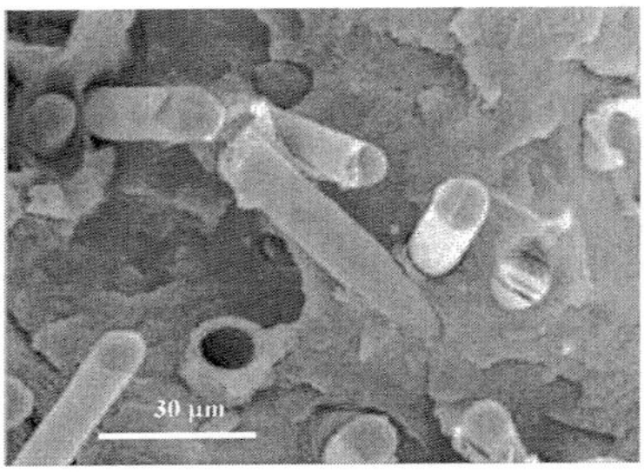

Figure 1. SEM micrograph of the fracture surface of PET-D: no adhesion between matrix and glass fibres.

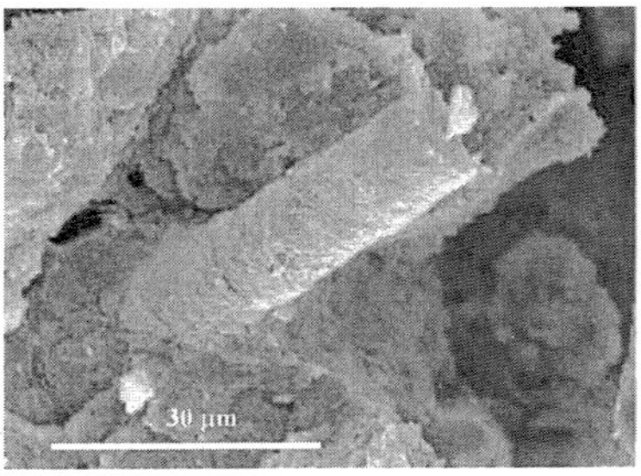

Figure 2. SEM micrograph of the fracture surface of PET-A: good interfacial adhesion between matrix and glass fibres.

Using as additives phosphorus-based halogen-free flame retardant and treated glass fibre, the first successful result was obtained for the material PET-C7, showing a V-0 rating. Next steps focused on optimising phosphorus and glass fibre contents.

A previous study [6] has observed that incorporation of glass fibres in a recycled PET in combination with a phosphorus-containing flame retardant significantly increases the stiffness of the char layer formed during combustion and limits the flow of the molten polymer. A condition to obtain this behaviour is good adhesion between the char layer and the glass fibres. The positive role of glass fibre can also be explained by the large amount of volatile combustibles that can be retained inside the char layer. This leads to the phenomenon of intumescence, which adds to the effect on fire resistance obtained by flame retardant, which contributes to create the char.

In table 1, phosphorus and glass fibre contents are shown in relative units related to phosphorus content of PET-C5, equal to 1.

Table 1. Relative composition, UL94 test and HDT of the studied materials.

Sample	P-cont.	GF-cont.	Burning time (s)	UL94 V-rating	HDT (°C)
PET-A	0.00	14.95	∞	V	>150
PET-B	0.00	0.00	29.1	V-2	65.5 ± 0.5
PET-C1	0.53	7.35	168.7	V	49.8 ± 1.1
PET-C2	0.55	9.95	162.4	V	50.0 ± 0.7
PET-C3	0.75	4.25	53.8	V-2	43.7 ± 2.1
PET-C4	0.85	10.80	93.2	V-2	42.9 ± 1.6
PET-C5	1.00	4.25	22.6	V-0	38.2 ± 0.5
PET-C6	1.09	10.10	32.3	V-0	37.7 ± 1.1
PET-C7	1.26	3.50	17.3	V-0	39.9 ± 0.1
PET-D	1.60	10.70	6.3	V-2	44.8 ± 0.2

HDT temperature decreases with increasing phosphorus content. The content of glass fibre does not seem to play a role, since pairs of values with a similar phosphorus content (and different glass fibre content) have a similar HDT temperature: PET-(C1/C2, C3/C4 and C5/C6).

Mechanical properties

In figures 3 and 4, results of impact tests are shown as a function of glass fibre and phosphorus content, respectively.

As expected, materials with treated glass fibre show better impact properties than the material with untreated glass fibre (PET-D). It is well known that adhesion of glass fibre to the matrix reduces propagation of cracks [7].

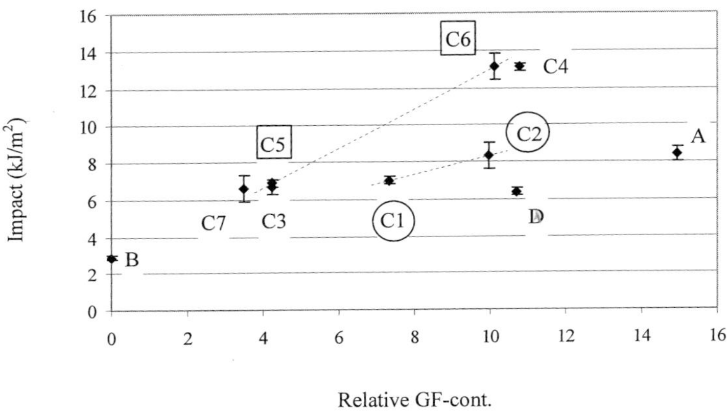

Figure 3. Impact strength of the studied materials as a function of relative glass fibre content. (Relative P-cont.: □ ≈ 1 and ○ ≈ 0.5).

For constant phosphorus content, a comparison of the materials PET-(C1/C2 and C5/C6) shows that increasing of glass fibre content leads to increasing of impact strength.

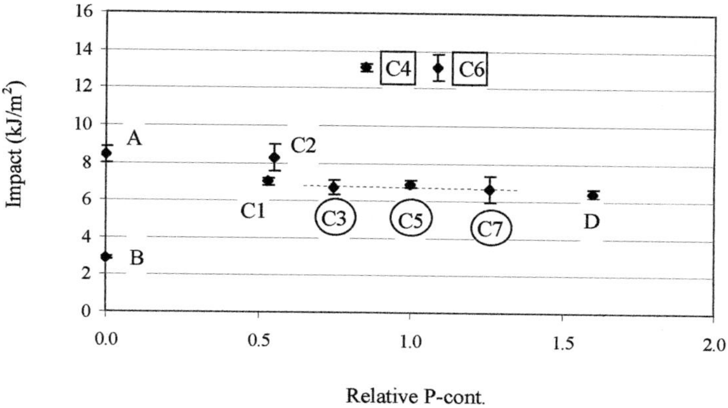

Figure 4. Impact strength of the studied materials as a function of relative phosphorus content. (Relative GF-cont.: □ ≈ 10 and ○ ≈ 4).

For constant glass fibre content, a comparison of the materials PET-(C4/C6 and C3/C5/C7) shows that impact strength is independent of phosphorus content. In the case of the materials PET-C2, a lower value is obtained, more tests should be performed to study this issue. Results obtained in the impact test are important to know the application range of the final product.

In figures 5 and 6, results of tensile tests are shown as a function of glass fibre and phosphorus content, respectively.

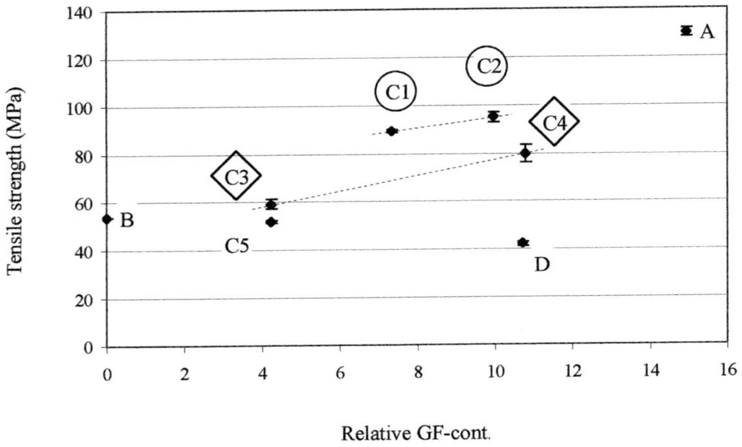

Figure 5. Tensile strength of the studied materials as a function of relative glass fibre content. (Relative P-cont.: ◊ ≈ 0.8 and ○ ≈ 0.5).

Material PET-D shows lower mechanical properties than PET-C# materials due to the lack of adhesion in the glass fibre as mentioned before. On the other hand material PET-A shows higher mechanical properties than PET-C# materials due to the high GF-cont with a good surface adhesion and because it is a virgin material (further research has shown that tensile strength of PET-A falls 23% after five recycling operations).

For constant phosphorus content, a comparison of the materials PET-(C1/C2 and C3/C4) shows that increasing of glass fibre content, leads to increasing of tensile strength.

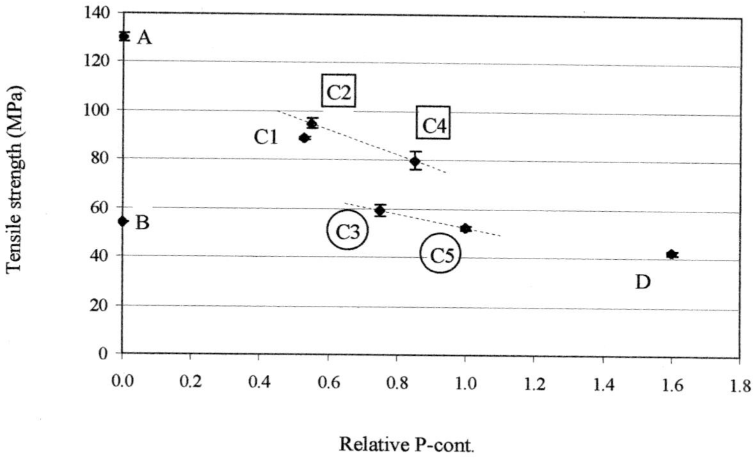

Figure 6. Tensile strength of the studied materials as a function of relative phosphorus content. (Relative GF-cont.: □ ≈ 10 and ○ ≈ 4).

For constant glass fibre content, a comparison of the materials PET-(C2/C4 and C3/C5) shows that increasing of phosphorus content leads to slight decrease of tensile strength.

Tensile tests also showed that Young Modulus, and therefore material stiffness, increases with increasing glass fibre fraction. This assessment was foreseeable, as the glass fibre is more rigid than the PET matrix.

Conclusions

A halogen-free recycled PET UL94 V-0 rating could be achieved with a combination of phosphorus-containing flame retardant and glass fibre.

The best way to avoid dripping in the flammability test of PET is the addition of glass fibre treated with a specific coupling agent for PET. Additionally, a compact structure is reached after the burning test.

Incorporation of a phosphorus-containing flame retardant neither seems to have an important influence on the mechanical properties, nor on the injection process. HDT temperature may be a limiting property for the incorporation of the flame retardant, but does not seem to be a disadvantage for the application of the reference material in TV components.

Further research could focus on optimising the synergistic effect of glass fibre and flame retardant with the assistance of an experimental design and analysis of the response surfaces obtained.

[1] Bandrup. et al.. *Recycling and Recovery of Plastics*. Hanser Publishers. **1996**. 669-675.
[2] H. E. Ferrando. J. F. Cribier. D. Vega. F. Bosch-Masgrau. M. Sanchez-Soto. M. Ll. Maspoch. *Polymer Recycling*. **2001**. 6(4). 187-192.
[3] C. S. Wang. J. Y. Shieh. Y. M. Sun. *Journal of Applied Polymer Science*. **1998**. 70. 1959-1964.
[4] A. Granzow. R. G. Ferrillo. A. Wilson. *Journal of Applied Polymer Science*. **1977**. 21. 1687-1697.
[5] S. J. Chang. Y. C. Sheen. R. S. Chang. F. C. Chang. *Polymer Degradation and Stability*. **1996**. 54. 365-371.
[6] F. Laoutid, L. Ferry, J.M. Lopez-Cuesta, A Crespy. *Polymer Degradation and Stability*. **2003**. 82. 357-363.
[7] J. A. Brydson. *Plastics materials*. 6th ed, ed. Butterworth-Heinemann. **1995**. 122-127.

Macromol. Symp. **2005**, *221*, 185-195

An Electroconductive Filler for Shielding Plastics

*B. Kandasubramanian, M. Gilbert**

Institute of Polymer Technology and Materials Engineering, Loughborough University, Loughborough, Leics, LE11 3TU, UK
E-mail: m.gilbert@lboro.ac.uk

Summary: A well characterised grade of mica was used to prepare nickel-coated mica fillers by an electroless coating technique. These fillers were incorporated into polypropylene (PP) and acrylonitrile-butadiene-styrene (ABS) using a Haake Internal mixer and a two-roll mill to produce conducting composites suitable for EMI shielding applications. Compounded polymers were fabricated into sheets of different thickness by compression moulding. The dependence of conductivity on sample thickness was studied and samples were tested for electrical resistivity, shielding efficiency, and thermal and mechanical properties. Reduced sample thickness during compression moulding decreased the electrical resistance of the polymer composites due to orientation and the formation of a good conducting network. Volume resistivity of PP composites was lower than for ABS composites with 50% Ni coating and equivalent filler weight fracions, showing that better conductivity could be achieved in a semi crystalline polymer than in an amorphous polymer. 0.6 weight fraction of 50 wt.% nickel coated mica in ABS showed a shielding efficiency of 16dB compared with 27.8dB in PP.

Keywords: ABS; electroless nickel coating; EMI shielding fillers; polypropylene

Introduction

Plastics capable of conducting electricity offer many advantages. They provide inherent EMI shielding to protect business machines, automotive and aerospace components and computers from electromagnetic interference. In addition to the consideration that a device should not emit radiation that interferes with others, electronic devices must be capable of functioning under EMI from other sources, i.e., they must exhibit electromagnetic compatibility, and must be shielded in such a way so that both incoming and outgoing interferences can be filtered. Plastics are widely used to make shells for electronic equipment. However, their volume resistivities are approximately 10^{13}-10^{17} ohm-cm, so they are transparent to electromagnetic waves.

A range of options is available for introducing conductivity to plastic housings and enclosures: these options broadly fall into three categories. Intrinsically conductive polymers such as polyacetylene, polyaniline, polythiophene and polypyrrole can be produced[1], but these are expensive, and difficult to process. Coating methods are the predominant choice for EMI shielding operations, accounting for 98% of the market in

DOI: 10.1002/masy.200550319

1995[2]. However, they have a number of drawbacks as they involve an extra production process, can be subject to delamination, and are difficult to recycle. The third approach is to produce plastic composites containing conductive fillers. In conductive polymer composite materials, a conductive filler is incorporated into a polymer matrix. These materials are relatively cheap and have good processibility. Also, the conductivity of a composite can be adjusted relatively easily to meet product requirements. The most common conductive fillers are carbon black, carbon fibres, metallic powders, flakes or powders, and fibres coated with metals [2].

The electrical resistivity of conductive composites depends on the volume fraction of the conducting filler particles, and is explained by the percolation theory. The conducting filler is required to produce a conductive particle network in the system. Electrical properties depend upon factors such as filler content, aspect ratio, conductivity of filler and the processing conditions. The amount of electrically conductive filler required to impart conductivity to an insulating polymer can be dramatically decreased by selective localisation of the filler in one phase, or better at the interface of a co-continuous two-phase polymer blend[3]. The critical amount of filler necessary to build up a continuous conductive network and accordingly to make the material conductive is referred to as the percolation threshold.

As indicated above, various conductive fillers have been used, but all have some disadvantages[1]. Effective conductive fillers such as nickel-coated carbon fibres are limited to specialist applications due to cost. In this work the use of coated mica has been investigated. Mica is a sheet silicate with a platelet structure having aspect ratios of 20 to 40. Due to its high modulus and flaky shape it is used in plastics to improve rigidity and dimensional stability[4].

If coated with metal, mica has the potential to retain dimensional stability and increase rigidity of the composite used, at the same time become a conductive filler. Properties of composites containing such a filler have been reported. Kortschot and Woodhams[5] emphasised the importance of aspect ratio, and suggested that this could be achieved by using a thin metal coating on a substrate such as mica. Low nickel concentrations could be used if thinner and more uniform coatings could be achieved. Du and Kumagani[6] mixed nickel-coated mica / epoxy and triethylenetetramine at 115 ^{0}C for 30 minutes. It was observed that nickel-coated mica flakes, oriented in the direction parallel to the sheet, which increased contacts between fillers, had a very good shielding effectiveness.

Previous work[7,8] has identified optimum conditions for the electroless coating of mica, and shown that conductive compounds can be produced using nickel coated mica. In this paper, coating has been carried out on a larger scale to enable composites to be produced.

Experimental

Materials

Mica TK 100C-100201, run number 595, phlogopite grade with layered profile supplied by Minelco Specialities Ltd., UK was used the investigation. The properties of mica are shown in Table 1. The compositional details of the mica are SiO_2 = 40 - 42%, Al_2O_3 = 9 - 11%, Fe_2O_3 = 9 - 11%, MgO = 22 - 24%, K_2O = 10 - 11%.

Table 1. Properties of mica.

Mica grade	Density (gcm^{-3})	Average particle size (μm)	Average particle thickness (μm)	Refractive index	Surface area (m^2g^{-1})	Average aspect ratio*
TK 100C	2.50	100	1	1.58	0.0613	37: 1

* Obtained from direct measurement of thickness (interferometry) and diameter (Carl Zeiss photpmeter) on about 200 particles

Electroless plating solutions used were Macuplex^R Activator D-34C, which is a liquid catalyst system containing both the sensitiser and the activator in a single system for pre-treatment of the specimens before electroless nickel or copper coating, and ELNIC 204, a stable electroless nickel solution. The process is used to deposit an ultra bright, low-medium phosphorous content coating on a variety of substrates. This solution consists of two process components ELNIC 204 A and ELNIC 204 B. ELNIC 204 A is the nickel component and ELNIC 204 B is the reducer and complexing component.

Polymers were used for this work were Escorene PP 7034 P1, a heterophasic polypropylene copolymer supplied in the form of granules by Exxonmobil with a melt flow rate of 12dg min^{-1} at 230 ^{0}C and 2.16kg, and the acrylonitrile-butadiene-styrene resin (Cycolac T ABS) from GE Plastics. Its composition is 25% acrylonitrile, 20% butadiene, 55% styrene with melt flow index of 26dg min^{-1} at 220 ^{0}C and 10kg load.

Chemistry of electroless coating process

This consists of three steps, described in detail previously[7,8].

Step 1. Sensitisation. Treatment with stannous chloride sensitises the surface of mica with Sn^{2+}.

Step 2. Activation. Sensitised mica particles were treated with a 2% solution of palladium chloride. Pd^{2+} is reduced to Pd and acts as the catalytic site for electroless nickel coating.
Step 3. Nickel coating. Pre-treated mica was then treated with different concentrations of nickel solutions based on the weight percent of nickel required. The coating was carried out between 85-90^{0}C, using a pH of 5.

Processing

PP, ABS and nickel coated mica were dried for two hours in an oven at 100^{0}C. These fillers were tumble blended with the resins. Compounding was done using a Haake internal mixer (screw speed 50rpm, mixing for 10 min at 210°C) and a two-roll mill (milling temperatures PP 170°C; ABS 180°C).
The compounds from the Haake mixer and two roll mill were compression moulded for 5 min to prepare 50 mm square test coupons of 1 mm thickness, unless otherwise stated. The moulding pressure was 5MPa; moulding temperatures were 230°C (PP) and 220°C (ABS).

Testing and Characterisation

Testing. Tensile properties were determined for samples cut from 1mm test sheets according to BS 903: Part A: 1971 using a Lloyd Tensile testing machine, type L200R fitted with a 500N load cell. The grip separation speed was 10mm min^{-1} at room temperature. The average of five results was reported. Flexural testing was carried out, according to ISO 178 with a test speed of 10mm min^{-1} using the same Lloyd tensile test machine. The Dapmat computer programme software calculated all flexural properties automatically. The average of five samples per formulation was reported.
Volume resistivity. The prepared test coupons were used to measure volume resistivity as described previously[8]. Resistivity was measured by a super-Megohmmeter type electrometer (Model RM 170) with a DC power supply source. When the applied voltage is small there is no current through the sample. As voltage increases there is a point (described as the break-through voltage of the sample) at which there is apparent current through the sample. One hundred volts, well above the breakthrough voltage of all samples in this study, was used for the measurement of volume resistivity. Reported values are the average values of the resistivity (ohm cm^{-1}) obtained from five measurements on each composite. For thinner samples, it was necessary to reduce the

applied voltage to 10V; at higher voltages the current was too high to be measured with the equipment used[7].

Shielding efficiency. The electromagnetic shielding efficiency of the polymer composites was carried out at QinetiQ Ltd., Farnborough. The sample geometry is shown in Fig 1. Samples were prepared by machining from compression moulded sheet. A washer shaped sample is electrically connected between the inner and outer conductor of a specially designed coaxial fixture, which is placed between the signal source and the detector. The ratio of the power of the signal received with the sample present to the power received with no sample is the shielding effectiveness of the sample. The sample must make good electrical contact at both the inner and outer surfaces, or the measured shielding effectiveness will be lower than the sample is capable of providing. The microwave transmission losses of the filled plastic materials within the range of 0.1-20 GHz, were measured at room temperature. A major drawback of this test is that it would not identify any property anisostropy which is common in filled polymer composites.

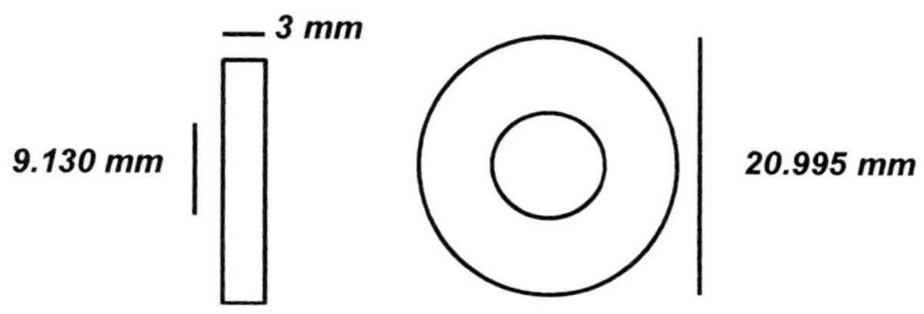

Fig. 1. Sample Dimensions for EMI test.

Results and discussion

Properties

Tensile and flexural data for the composites produced are shown in Tables 2 and 3. Both tensile strength and elongation decrease with increase in filler loading, but also as the % coating increases. No reinforcement is being provided by the mica as observed by Yang et al[9] who incorporated nickel-coated carbon fibre into ABS. The decrease in tensile strength as the amount of nickel coating increases suggests that the nickel is impeding polymer filler interaction. It has been demonstrated that in filled composites[10] stress concentration points are introduced through dewetting of the filler from the polymer and these will create discontinuities in the structure rendering it weak.

Flexural modulus increases with filler concentration due to the introduction of a highjer modulus filler as expected, but also slightly as the amount of nickel coating increases, presumably because of the high modulus of the nickel coating itself. The effect of the coated nickel has been investigated, and will be reported in a future publication.

Table 2. Mechanical properties of polypropylene composites.

Filler loading in composite (Wt.Frac.)	% Ni coating on mica	Tensile Strength (MPa)	Elongation (%)	Flexural modulus (MPa)
PP	-	30.8	45.5	679
0.1	30	27.3	22.5	869
0.2	30	22.8	13	1050
0.3	30	19.9	10.5	1188
0.4	30	15.5	4	1276
0.5	30	11.1	3.5	1401
0.6	30	9.4	3	1646
0.1	40	22.8	20.5	1161
0.2	40	19.6	11	1385
0.3	40	17.9	6	1629
0.4	40	15.3	3.5	1840
0.5	40	10.1	3	1969
0.6	40	8.9	2.5	2021
0.4	45	13.8	3	1888
0.5	45	9.5	2.5	1989
0.6	45	7.9	2.5	2086
0.4	50	10.2	2.5	1965
0.5	50	7.2	2	2123
0.6	50	5.9	1.5	2145

Volume resistivity

Table 4 and Fig 2 show volume resistivities for 1mm thick composite samples produced using a pressure of 5 tonnes. The volume resistivity of composites containing conductive filler depends on the proximity of neighbouring conductive particles. Table 4 shows that volume resistivity depends on both the amount of nickel on the mica, and the weight fraction of the coated filler. No significant change in resistivity was observed when the mica was coated with 30% nickel.

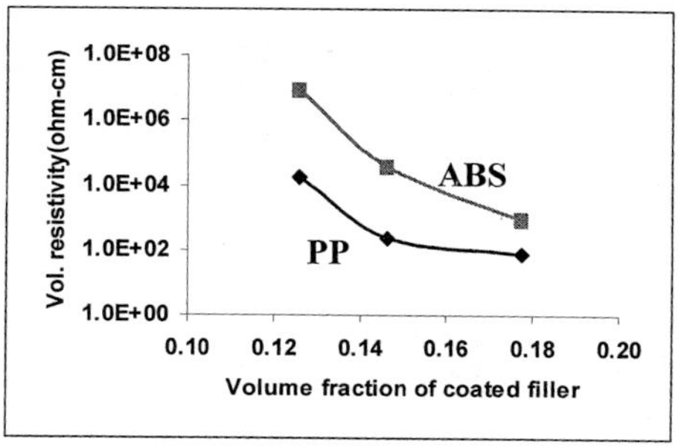

Fig 2. Comparison of resistivities of PP and ABS.

Table 3. Mechanical properties of ABS composites.

Filler loading in composite (Wt.Frac.)	% Ni coating on mica	Tensile Strength (MPa)	Flexural modulus (MPa)
0	-	51.6	1742
0.4	50	31.5	2458
0.5	50	16.2	5532
0.6	50	12.1	7481

Table 4. Volume resistivities of polypropylene containing nickel coated mica.

Filler loading in composite (Wt.Frac.)	% Ni coating on mica	Volume resistivity (ohm-cm)
PP	-	10^{16}
0.1	30	8.72×10^{11}
0.2	30	7.24×10^{11}
0.3	30	7.03×10^{11}
0.4	30	6.86×10^{11}
0.5	30	6.51×10^{11}
0.6	30	6.23×10^{11}
0.1	40	6.89×10^{11}
0.2	40	5.62×10^{11}
0.3	40	4.26×10^{11}

Filler loading in composite (Wt.Frac.)	% Ni coating on mica	Volume resistivity (ohm-cm)
0.4	40	3.66×10^{9}
0.5	40	2.90×10^{9}
0.6	40	9.55×10^{8}
0.4	45	8.62×10^{8}
0.5	45	6.91×10^{8}
0.6	45	7.55×10^{6}
0.4	50	5.62×10^{4}
0.5	50	6.91×10^{2}
0.6	50	9.60×10^{1}

At higher Ni levels, resistivity started to decrease as filler loading increased. At very low filler loading, the resistivity of the composites is nearly equal to that of the base polymer as the conductive particles are surrounded by the insulating polymer. The resistivity starts to fall rapidly when the nickel-coated mica content reaches a particular value, known as the percolation threshold. After this transition, the rate of change of resistivity with filler concentration will reduce. At the percolation point continuous contacts are made between adjacent particles to form conductive network, which facilitate the electrical conduction through the composites. In order to obtain conductive composites, the filler itself must be conductive, so that an adequate amount of nickel should be present. Mica coated with ≥ 40% Ni had a volume resistivity of approximately 1 ohm-cm. Clearly a resistivity of this order is necessary for it to function as a conductive filler. The performance of nickel-coated mica flake with 50 wt% Ni was almost identical to that of commercial aluminium flakes. This is expected since the aspect ratio of the coated mica is quite similar to that of aluminium flakes, both being about 20 to 50[5,11].

Fig 2 compares data for PP and ABS at equivalent volume fractions of filler. In semi-crystalline polymers small fillers can be segregated into amorphous regions. In such polymers the critical concentration required to achieve network formation is less than in a purely amorphous polymer where uniform dispersion is achieved. The PP composites based on PP have lower resistivities, suggesting that in this case a conductive network is formed by the concentration of filler in amorphous regions, while in amorphous ABS the nickel coated mica will be distributed throughout the polymer matrix as explained by Reboul et al[12]. The effect of sample thickness was also investigated, and results are shown in Table 5. It is seen that resistivity decreases as samples are pressed to make thinner sheets, due to improved contact between the mica plates.

Table 5. Effect of sample thickness on volume resistivity of PP composites containing 0.6 weight fraction of 60% Ni coated mica.

Sample thickness (mm)	0.2	0.3	0.5	1.0
Volume resistivity (ohm-cm)	4.1×10^1	5.6×10^1	7.2×10^1	9.6×10^1

Volume resistivities of composites of PP and 40% nickel-coated mica fabricated by two different techniques are shown in Table 6. The samples fabricated using a two-roll mill showed better conductivity than samples produced with the Haake mixer. Figures 3 and 4 compare the appearance of the composite with 40% nickel-coated mica using different processing techniques. It is seen that the partcle size of nickel-coated mica was reduced when the composite was compounded in a Haake batch mixer due to the higher shear experienced, but more larger particles with higher aspect ratio are present after mixing on a two roll mill, explaining the better conductivity of this compound[7]. The average particle size of ashed polymer composite processed in the Haake batch mixer was 37.5 μm compared to 75.1 μm for the filler compounded in a two roll mill. Micrographs showed no evidence of adhesion between the mica flakes and the PP matrix.

Table 6. Effect of processing technique on volume resistivity of PP composites containing 40% Ni coated mica.

Weight fraction of filler	Volume resistivity (ohm-cm)	
	Haake	Two roll mill
0.4	4.01×10^9	3.21×10^9
0.5	3.96×10^9	3.11×10^9
0.6	9.97×10^8	1.71×10^8

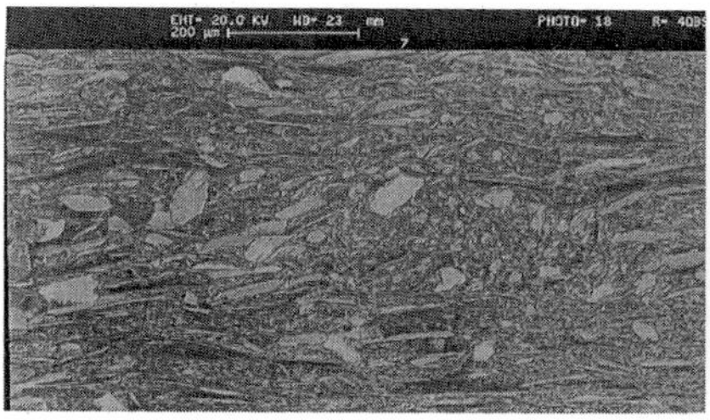

Fig. 3. PPcompounded with Ni coated mica in Haake.

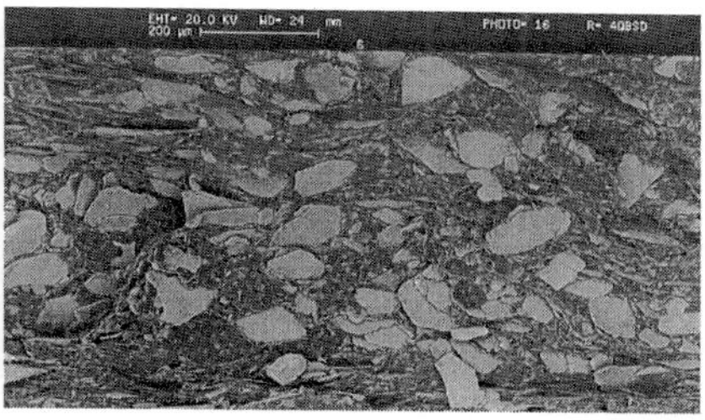

Fig. 4. PP compounded with Ni coated mica using two roll mill.

Shielding efficiency

Shielding efficiency data for samples with the lowest resistivities are shown in Figure 5. The test samples are assumed to be isotropic and homogeneous. The shielding effectiveness of 0.6 weight fraction of 40% nickel coated mica with PP was found to be 9.1dB for the frequency range of 750 – 2000MHz. The value was found to be consistent across this frequency range. 0.6 weight fraction of 45% nickel coated mica had shielding efficiency of 10.1dB. This result was consistent between 500 to 2000MHz. 0.6 weight fraction of 50% nickel coated mica showed the maximum value of shielding efficiency of 27.8 dB at 1000MHz, but, there was a variation observed as a factor of frequency. In line with the resistivity data, the equivalent ABS composite had a lower shielding efficiency of 16dB.

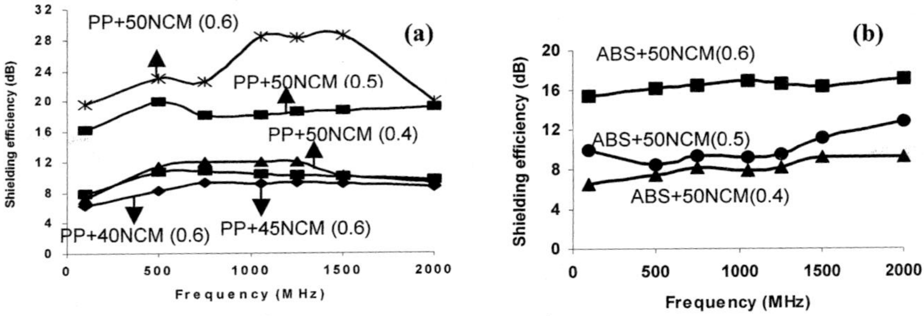

Fig. 5. Shielding effectiveness of composites (a) polypropylene; (b) ABS.

The effect of processing method is illustrated in Figure 6. The composite produced using the two-roll mill had a shielding efficiency of 9.1dB when compared to 7.5dB for the Haake mixed system prepared with different processing techniques The reduction in the particle size of nickel-coated mica clearly reduces the effectiveness.

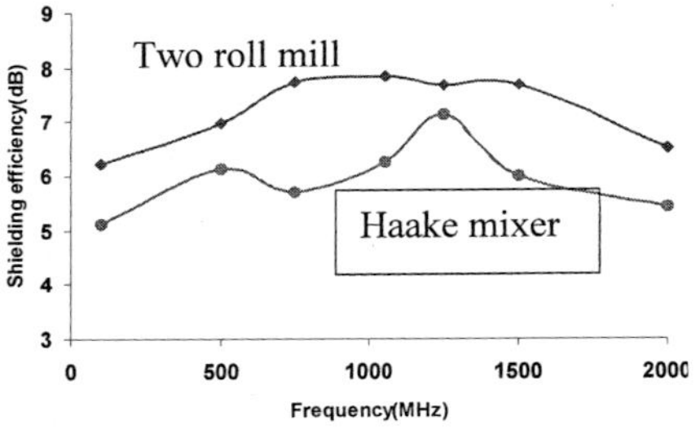

Fig. 6. Shielding effectiveness for composites processed using different techniques.

Conclusions

Tensile properties deteriorated when nickel coated mica was added to PP and ABS, while modulus increased as expected. Micrographs suggested that the interfacial adhesion between the polymer and nickel coated mica is poor, which will explain the poor properties. Conductivity increased with both filler concentration, and the amount of Ni on the mica. 40 – 50 wt % of coating is necessary to produce conductive compounds. Better conductivity and shielding efficiency was achieved in PP than in ABS, probably because the filler is confined to amorphous regions so that a conductive path is established more easily. Conductivity increased as sample thickness decreased because of improved contact between the conductive flakes. Compounds produced on a two roll mill had lower resistivity than those using a Haake rheometer; in the latter case the filler underwent attrition because of the higher shear forces during mixing. The highest shielding efficiency of obtained was 27.8dB for the PP based composite with 0.6 weight fraction of 50% nickel coated mica. The volume fraction of Ni in this composite was 0.06, so conductivity had been achieved at a low nickel concentration.

[1] D. Bigg, *Electrical properties of composites*, Marcel Dekker Inc., NewYork, **1986**, 197.
[2] Electromagnetic interference shielding, *A material prespective, Innovation 128, Tech trends*, International report on advance technologies, **1996**.
[3] K. Cheah, M. Forsyth and G. P. Simon, *J. Polym. Phys.: Polym. Phys.*, **2000,** *13*, 3106.
[4] H.Zweifel, *Plastics Additives Handbook*, Hanser Gardner publications Inc., Cincinnati, **2000,** 928.
[5] M.T.Korschot and R.T.Woodhams, *Polym. Comp.*, **1985,** *6,* 296.
[6] J.Du and Y.Kumagai, *J.Mater.Sci. Lett.*, **1994,** *13*, 1786.
[7] M.Gilbert and B. Kandasubramanian, *Conductive compounds, Macromolecular Symposia*, **2003,** *194*, 219.
[8] G. Jiang, M.Gilbert, D. J. Hitt, G.D.Wilcox, K.Balasubramanian, *Comp: Part A,* **2002,** *33*, 745.
[9] S. Y.Yang, C. Y.Chen and S. H. Parg, *Poly. Comp.*, **2002,** *23*, 1003.
[10] J. Karger-Kocsis, *Polypropylene Structure, blends and composites*, Volume 3, Chapman & Hall, London, **1995**.
[11] M. T. Kortschot and R. T. Woodhams, *Polym. Comp.*, **1988,** *9*, 60.
[12] J. P. Reboul and G. Moussalli, *Int. J. Polym. Mater.*, **1976,** *5*, 133.

Models for the Investigation of the Structure of Vulcanizates

Jacek Magryta

Instytut Przemysłu Gumowego "STOMIL", ul. Harcerska 30, 05-820 Piastów, Poland
Fax:(+48)0 227237196; E-mail:j.magryta@ipgum.pl

Summary: For the investigation of vulcanizates structure, new models have been developed relating the modulus of vulcanizates and the volume fraction of filler. The parameters of the models are the filler distribution and the properties of the shell surrounding the filler particles. Three functions of filler distribution have been assumed. Dynamic mechanical properties of a series of vulcanizates were tested and the parameters of the models were computed. The obtained data allows estimating the properties of the shell formed around filler particles in vulcanizates and to distinguish the functions of filler distribution.

Keywords: carbon black; elastomers; fillers; modeling; reinforcement

Introduction

Carbon black added to a polymer system causes a considerable change in its properties, particularly in the dynamic properties of vulcanisates. Understanding of the mechanism of reinforcement of elastomers by carbon black is one of the most important problems faced by rubber science and technology. It is known that the polymer - filler interaction leads to the formation of rubber shells around the filler particles. The modulus of the shells formed around carbon black particles in vulcanizates is higher than the modulus of the rubber matrix.

The properties of this shell have a great influence on the properties of rubber (e.g. dynamic modulus) [1-3] and can be a clue towards explaining the phenomenon of reinforcement. One of the possible ways of looking for an explanation of this phenomenon and the behaviour of filled vulcanizates is to develop models assuming the existence of particular structures producing the changes in the properties. If a good agreement between the experimental and the theoretical data is obtained then an analysis of parameters of a model describing the structures can be performed.

The effects of fillers on dynamic mechanical properties of rubber give some important information about the structure of vulcanizates. Therefore, this paper will discuss the models derived to describe the dynamic mechanical properties of rubber in order to investigate the structure of vulcanizates.

 DOI: 10.1002/masy.200550320

Theory

It is well known, that the time of a mixing a rubber compound has a great influence on the level of dispersion of carbon black and on the dynamic modulus of the rubber [4-5]. Even from such a simple experiment it is obvious that the dispersion of filler is an important factor. However, the importance of how it relates to others parameters can be examined by models.

Rubber - filler interaction is represented in this model by the value of the modulus and dimension of the shells existing around particles of carbon black.

A multiphase cuboid system is divided into equal cubes containing different amounts of filler placed at corners (Figure 1) [6].

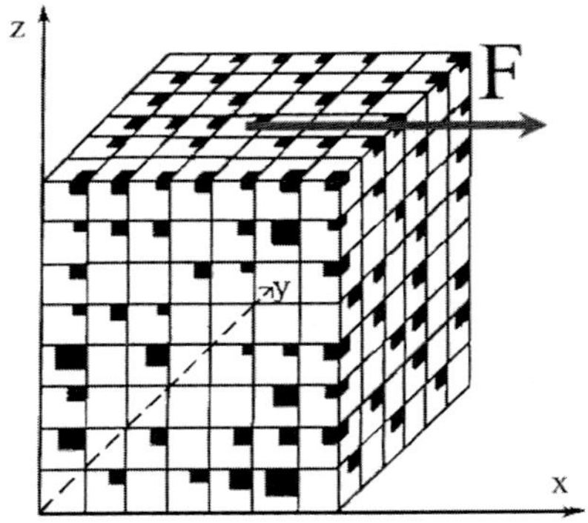

Figure 1. Three dimensional model with random distribution of carbon black.

Each of the cubes can be represented by the models described earlier [7]. Good adhesion and incompressibility of each phase have been assumed. A simple shear force has been applied to the cuboid but the forces and deformation of the individual cubes are not known a priori.

The forces carried by the planes parallel to X-Y have the same value:

$$F = \sum_{i=l}^{i=l} \sum_{j=1}^{j=n} F_{i,j,k} \tag{1}$$

where: $F_{i,j,k}$ is the force acting on the cube with the coordinates i,j,k and l and n are the number of cubes on the axes x and y.

Displacement d of the upper surface relative to the lower is equal to the displacement of a column for which the indices i and j are fixed, leading to:

$$d = \sum_{k=1}^{k=m} d_{i,j,k} \tag{2}$$

Using the definition of modulus (G=τ/γ) and counting the sums of cubes in the model, the following formula is obtained:

$$G = \frac{mF}{D\sum_{i=1}^{i=l}\sum_{j=1}^{j=m}\sum_{k=1}^{k=n} d_{i,j,k}} = \frac{mF}{D\sum_{j=1}^{j=N} d_j} \tag{3}$$

where: D is the dimension of the elementary cube, and N the number of all cubes in the model (N=l*m*k) .

As the result of the calculation with some simplification and grouping of expressions, the following expression has been obtained:

$$G = \frac{1}{\sum_{j=1}^{j=k}\left(\frac{a_j n_j}{G_j N}\right)} \tag{4}$$

Where: G-modulus of system; a_j - factor characterizing the stress concentration, n_j – number of the cubes containing the filler concentration in the ranged c,c+dc; G_j-modulus of "j" cube and N number of the cubes in the model.

The expression n_i/N is the probability of appearance of a cube with a volume fraction of filler between c and c+dc. It is also the probability of the occurrence of a_j and G_j, if they are monotonous functions of c.

Going up to an infinite number of cubes, the sum is replaced by integration:

$$G = \frac{1}{\int_0^1 \frac{a(c)}{G(c)} f(c)dc} \tag{5}$$

Three functions of the model are unknown and should be fixed:

- G(c) –modulus of the elementary cube,
- f(c) - function of probability of the density of filler distribution,
- a(c) -stress concentration function.

Based on existing works it is difficult to propose a density function of the probability of filler distribution in rubber. Therefore, three types of distribution were considered:

- parabolic distribution (approximation of Gaussian function)
- uniform distribution
- inverted parabolic distribution

There are two parameters describing the function of filler distribution, namely: **a** and **b** (parameters of functions, which are the lowest and the highest possible values of filler amount) or $\bar{y}$ and $\sigma = s^2$ (estimators of functions which are the average value of the filler fraction and the standard deviation of filler distribution).

The function of stress concentration indicates the ratio of the force acting on an elementary cube to the average value of the force. The preliminary calculation showed that a sensible function is a certain power of volume fraction of filler, e.g. $(\mathbf{a(y)=C(a,b)y^3}$ where $\mathbf{y=c^{1/3}})$.
The moduli of elementary cubes are calculated from assuming the appearance of a shell of rubber matrix on the filler surface. The shell is a result of rubber-filler interaction because the adsorption of polymer molecule chains on the filler surface may reduce their mobility)[2,7].
Each of the cubes can be represented by the models described earlier (model B)[7]. The properties of the shell (thickness and modulus) are the parameters of the model.

In order to become a general problem solving tool, the model should be dimensionless and the parameters should have a relative meaning.
The whole model has three parameters viz. dimension, r (ratio of the thickness of the shell to the side of the elementary cube D) and relative value of the modulus of the shell, m (ratio of the modulus of the rubber matrix, G_o, to the shell modulus, G_{shell}) and the parameter of the distribution of filler[7].

Application of the Theory

The tested samples were sulphur vulcanizates of NR and Cl-IIR as well as blends of both rubbers with N990 and N220 carbon black. For the vulcanizates of NR filled with N220 carbon black, different times of mixing of these compounds were applied to obtain different dispersions of the fillers. The concentration of carbon black for both types of compounds ranged from 0 to 80phr. The dynamic properties of the vulcanizates were determined using a D-8 apparatus produced by H.W. Wallace of Croydon. Simple shear forces with the frequency of 0.25Hz were applied. The test results were obtained in a range of amplitudes of deformation from $\gamma = 0.004$ to $\gamma = 0.35$.

The model parameters were calculated until good agreement between theoretical and experimental data was obtained. The goodness of fit was checked by the χ^2 test.

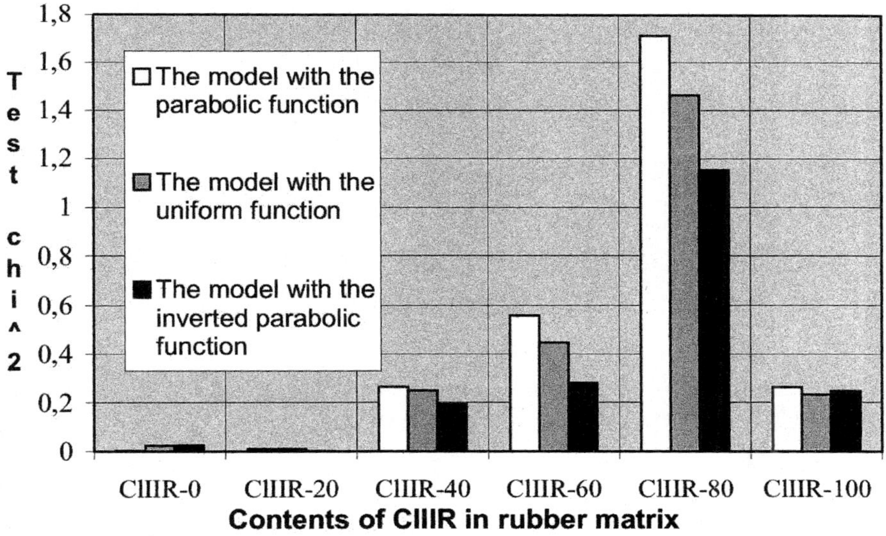

Figure 2. Test of goodness of fit of model-to-experiments data for NR+Cl-IIR +N990 carbon black vulcanizates at the amplitude of shear deformation $\gamma = 0.004$.

The results show good agreement with the experimental data. The values of the test of fitness in most cases are much better than the critical value for the level of confidence of 0.9. These good values of agreement confirm the reasonability of the model and allow discussion of the structure parameters. The analysis of values of the χ^2 test leads to the suggestion that the inverted parabolic function of the filler distribution is the best for the tested vulcanizates (Figure 2-3). The average values of the test χ^2 for the vulcanizates with N990 carbon black are 0.242, 0.209, and 0.166 for parabolic, uniform and inverted parabolic function of the density of probability, respectively.

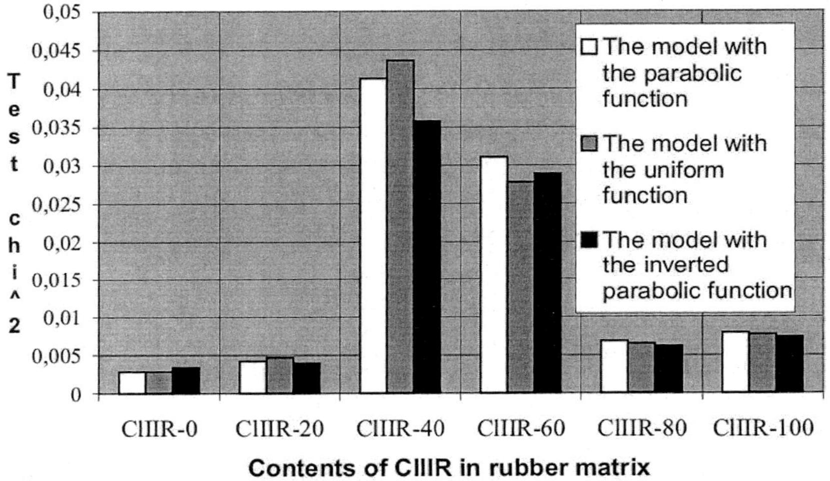

Figure 3. Test of goodness of fit of model-to-experiments data for NR+Cl-IIR +N990 carbon black vulcanizates at the amplitude of shear deformation $\gamma = 0.35$

The average values of the χ^2 test for the vulcanizates of NR compounds/N220 carbon black with different times of mixing are on a similar level, namely: 0.825, 0.711, and 0.438 for the three functions of filler distribution respectively as shown in Table 1.

The parameters s^2 (the percentage of maximum variance[7] of the probability functions of the filler distribution) become smaller when the amplitude of deformation increases for vulcanizates filled with N990 and N220. This can be explained as a result of breaking out the large filler aggregates or agglomerates (probably temporarily) and destruction of the filler-filler network [4].

The value of s^2 for the smallest amplitude of deformation is very high. The tendency to reach such a high value indicates the existence of large aggregates of filler, which carry over the force from one cube to another. Unfortunately, such phenomenon is not directly considered in the proposed models.

The parameter of the shell thickness allows to calculate the volume fraction of the shell. The calculated values strongly depend on the type of the functions of the density of probability and range thus:

- parabolic function, from 1,5 to 25.9%,
- uniform function, from 2.2 to 27.5%,
- inverted parabolic function, from 1 to 3.7%,

when vulcanizates are filled with 80phr of N990 carbon black.

As can be seen in Figures 2 and 3 for the low structure carbon black (N990), the data for the inverted parabolic function seems to be more accurate; the same is true for high structure carbon black (N220). The parameters of modulus of the shell should be higher than modulus of the rubber matrix. However for the vulcanizates filled with non-reinforcing black N990 (and the large deformation) in the majority of cases (except 60 and 80 % of chlorobutyl rubber in the rubber matrix) the value of the modulus of the shell is lower than for the rubber matrix. This phenomenon has been pointed out before (called eye-ball shape separation) and has been explained as a result of voids produced by the separation of carbon black particles from rubber matrix [7-8]. The separation is due to the deformation of the material and indicates low adhesion between filler particles and rubber.

In the vulcanizates with 60 and 80% of Cl-IIR presumably unequal distribution of carbon black in both rubbers lead to formation of quite strong filler network which behave very similar to networks of active fillers.

The moduli of the shells (the m parameters) calculated for the vulcanizates with N220 carbon black are much higher than the modulus of the rubber matrix (Table 1). Therefore, in this case separation of the rubber matrix from the surface of carbon black is not observed. As can be expected, if the time of mixing of rubber compounds is longer then the parameter of distribution is lower, which shows better dispersion of the filler. Also the parameter r (thickness of the shell) becomes lower when the time of mixing is longer. A better dispersion of the filler and a greater surface accessible to an interaction with the rubber matrix can give this result.

Table 1. Models with parabolic, uniform and inverted parabolic functions of filler distribution, results for NR + N220 compounds with different times of mixing.

Parabolic

Sample	Deformation amplitude	Parameters of models			Test of goodness of fit; χ^2
		s^2	r	$m(G_o/G_{shell}) * 10^{-6}$	
A (2min)	0.004 0.35	27 2.9	0.215 0.093	12 80	1.577 0.141
B(5min)	0.004 0.35	27 0.1	0.211 0.092	14 63	2.271 0.0376
C(10min	0.004 0.35	27 0.1	0.209 0.095	15 44	1.866 0.824
D (15min)	0.004 0.35	27 0.1	0.210 0.091	16 47	1.685 0.0474
E (20min)	0.004 0.35	27 0.2	0.189 0.087	18 39	1.429 0.0235

Uniform

A (2min)	0.004 0.35	25.5 0.1	0.174 0.102	10.7 48	1.390 0.131
B(5min)	0.004 0.35	27 0.1	0.164 0.091	10.6 52	1.961 0.037
C(10min	0.004 0.35	26.9 0.1	0.162 0.093	10 54	1.402 0.81
D (15min)	0.004 0.35	26.8 0.1	0.161 0.090	10.1 8	1.51 0.047
E (20min)	0.004 0.35	27 0.2	0.152 0.095	10.2 43	1.217 0.023

Inverted parabolic

A (2min)	0.004 0.35	27 9.8	0.081 0.0065	200 2100	0.923 0.195
B(5min)	0.004 0.35	27 8.4	0.077 0.0085	480 1300	1.284 0.0547
C(10min	0.004 0.35	26.9 8.3	0.065 0.0109	240 750	0.755 0.0547
D (15min)	0.004 0.35	26.8 7.8	0.075 0.0125	190 590	1.026 0.060
E (20min)	0.004 0.35	27 8.4	0.076 0.0140	190 250	0.8182 0.0314

For all distribution functions of filler (the three functions mentioned above) the higher values of modulus were obtained for the smallest amplitude of deformation. The most sensible value of the modulus of the shell for the inverted function of filler distribution, ranged from 100 to 1000 times higher than the modulus of the rubber matrix. These results are in good agreement with the published results [8, 9]. The values of r parameters obtained for the inverted function have the lowest value compare to other functions; they are also the most reasonable values.

The modulus parameters calculated for different functions of filler distribution do not vary significantly and all of them change in the same way with the content of chlorobutyl rubber in the rubber matrix. For example, it was found that all functions give the smallest modulus of the shell at the same composition of the rubber matrix.

From the comparison of the results obtained for the models assuming variability of the shape factor of filler particles[7] with the model considered here, it seems obvious that the meaning of these elements of structure is similar. In both cases, similar values of test of goodness of fit were obtained (Figure 4).

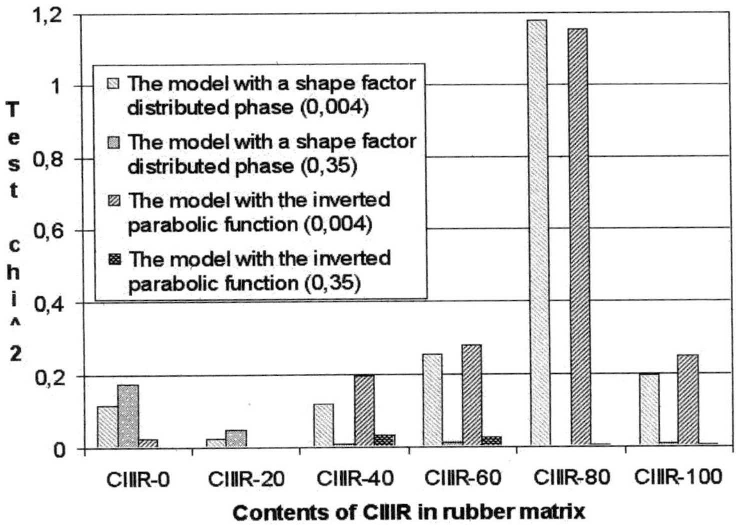

Figure 4. Comparison of data (NR+Cl-IIR +N990) of the model with the inverted parabolic function of filler distribution, with the model assuming a shape factor of filler particles.

Conclusions

All agree that the filler as one of the main component of the filled vulcanizates has a very important role in improving the dynamic performances of the rubber products. Many new ideas, theories, models and observations about how and why the filler alters the properties of vulcanizates have been presented. This suggests that multiple mechanisms of these phenomena may be involved. The properties of the shell formed on the filler surface as a result of polymer-filler interaction play an important role in the explanation of the modulus change of the rubber systems and other reinforcing effects [2,3,7,8].

The search for better understanding of the structure of vulcanizates and description of the properties vulcanizates called for the examination of the influence of the filler distribution in the rubber matrix. The distribution seems to be an important factor in the structure of vulcanizates. The presented results confirm this view. After investigating the models, it is quite clear that by substituting different functions of filler distribution to the model, it is possible to obtain various influence of volume fraction of filler in the modulus of a rubber system. The data obtained indicates that the influence of filler distribution on the dynamic modulus of vulcanizates is on a similar level as the shape factor of filler particles.

Three simple functions of density of the probability of filler distribution were proposed. The function of inverted parabolic seems to be the most interesting. If this function is really more common, therefore it would indicate that it is more probable to find spaces with huge concentration of filler and the spaces with very low concentration than the spaces with average concentration of filler.

The values of the test of goodness of fit are much better in most cases than the critical value (1,61) for the level of confidence of 0.9, meaning that the agreement of model to experimental data is quite good.

The data obtained (presented here and in [6]) makes it possible to draw a number of conclusions, some of them are indicated below:

- The distribution of filler seems to be an important factor in vulcanizates' structure and can help to explain the reinforcing effects.

- The parameters „S^2 (percentage of the maximum variance) become smaller when the amplitude of deformation increases, it can indicate that during large deformation the filler-filler network or the agglomerates of carbon black may break down.
- The most probable distribution of filler is the inverted parabolic function. The modulus of the shell formed around filler particles is smaller (except for 60 and 80% of CI-IR) than the calculated modulus for the rubber matrix for vulcanisates with N990 carbon black; for N220 the moduli of the shell are approximately 500 times higher than modulus of matrix.
- The thicknesses of the shell (parameter r) become lower as the time of mixing of the rubber compounds are longer.
- The parameter r allows calculating the volume fraction of immobilised shell, ranging from 1% to 26% of the rubber matrix.

Acknowledgments

This work is supported by the State Committee for Scientific Research (Poland) as research project in the years 2002-2005.

The author thanks the coworkers from "STOMIL" Rubber Research Institute for support and stimulating discussions.

[1] J. B. Donnet, Rubber Chem. Technol. 1998, 71(3), 323-341.
[2] M-J Wang, Rubber Chem. Technol. 1998, 71(3), 520-58.
[3] A.Vidal, B. Haidar, Elastomery (Poland), 1999, 3, 28-33.
[4] A.R. Payne, J. Appl. Polym. Sci., 1965, 9, 2273-2284.
[5] J. Magryta, R. Gaczyński, 1982, Polimery (Poland), 27,479-482.
[6] J. Magryta, Eur. Polym. J., 1991, 27, (4/5), 359-364.
[7] J. Magryta, R. Gaczyński, Eur. Polym. J., 1983, 19, (3), 205-212.
[8] Y. Fukahori, Rubber Chem. Technol, 2003, 76, 548-565
[9] Y. Fukahori, W. Seki, J. Mater. Sci., 1993, 28, 4471

Macromol. Symp. **2005**, *221*, 209-216

Predictive Models of Ethylene-Vinyl Acetate (EVA) Copolymers with Powdered Zn Fillers

Rafael Balart, Francisco Parres, Juan López, David García*

Higher Polytechnic School of Alcoy (Polytechnic University of Valencia), Paseo del Viaducto, 1, 03801, Alcoy, Spain
E-mail: rbalart@mcm.upv.es

Summary: Metallic Zn filler was used to increase the electrical properties of ethylene-vinyl acetate copolymer (EVA). Other properties were also modified such as mechanical and thermal behavior. It is possible to predict the behaviour of EVA-Zn system by using different models.

Keywords: composites; electrical properties; fillers; mechanical properties

Introduction

Ethylene vinyl-acetate copolymers (EVA) are broadly used in the industry (toys, footwear,...) due to a good combination of mechanical properties (elasticity) and easy processing. There are many and varied materials that can be incorporated to these copolymers with certain purposes: coloration, stabilization, foaming...

There are many works devoted to the development of polymers with good electrical properties for use in the production, for example, of composite electrodes[1-2]. The incorporation of metallic fillers represents an interesting procedure to modify the electrical properties of polymers [3].

We have worked on the development of composite electrodes that use EVA as matrix with the incorporation of powdered metallic Zn as filler. Mechanical and thermal properties have been studied to establish the features and the range of use of this material.

On the other hand, we have worked on the development of predictive models that help us to predict the properties of these materials. The use of general theories such as percolation theory has given excellent results in the prediction of electrical properties of conductive polymers.

We have made the use of this theory to the mechanical and thermal properties in order to establish the limitations of the theory and optimize the proposed models. Determination of

 DOI: 10.1002/masy.200550321

percolation thresholds of the different properties can help us to establish the validity of this theory.

Experimental

Materials

i) EVA copolymer. We have used EVA ALCUDIA ® PA-420 supplied by REPSOL CHEMICAL, INC. in pellet form, which has been subjected to a pulverization process in order to make easier the mixture with metallic Zn powder. This it is characterized for its flexibility and it is specially suitable for injection moulding. It contains an antioxidant as additive.

ii) Zn powder. Fine powder of metallic Zn supplied by MERCK has been used. EVA and Zn powders were mixed by a dry mixing process with vibration to improve the homogeneity of the mixture.

Sample preparation

Plates with dimensions of 120x60x4 mm were obtained with different amounts of Zn powder (10, 20, 30, 40, 50, 60, 62.5, 65, 70, 72.5, 75% Zn by weight) by means of the use of a hot plate press. These plates were prepared with a constant pressure of 10 kg/cm^2 during 180 s at 120 °C and later a cooling process to 40 °C for 600 s was applied.

Methods and characterization techniques

We have prepared samples according to the procedure established in ISO-527. Tensile tests of specimen were carried out in a universal test machine IBERTEST ELIB 30 (S.A.E. Ibertest, Madrid, Spain) at 10 mm/min.

Thermogravimetric analysis of EVA copolymer was carried out in a METTLER TOLEDO TGA-SDTA 381e (Mettler Toledo Inc., Schwerzenbach, Switzerland) in order to determine the content of vinyl acetate in the copolymer. Samples were treated under the following conditions : flow of N_2 (20 ml/min), weight of the sample (8-10 mg), ramp of temperature (30-700 °C at 10 °C/min). To determine the best processing conditions, calorimetric studies of different compositions have been carried out with a METTLER TOLEDO DSC STAR-2000 (Mettler Toledo Inc.,

Schwerzenbach, Switzerland). Samples were treated under the following conditions : flow of N_2 (50 ml/min), weight of the sample (5-10 mg), ramp of temperature (30 – 450 °C at 10 °C min).

Results and Discussion

Characterization of EVA-Zn system

Thermogravimetric analysis allows to determine the percentage of vinyl acetate in the EVA copolymer (Figure 1). It is possible to determine this content because degradation of EVA copolymers occurs in two steps. The first one is attributed to vinyl acetate phase degradation. This content results to be approximately of 20 %wt. It is important to know the percentage of vinyl acetate because of its polar behavior which can contribute together with Zn powder to change the electrical properties. This content is the typical for applications in toys and footwear industry.

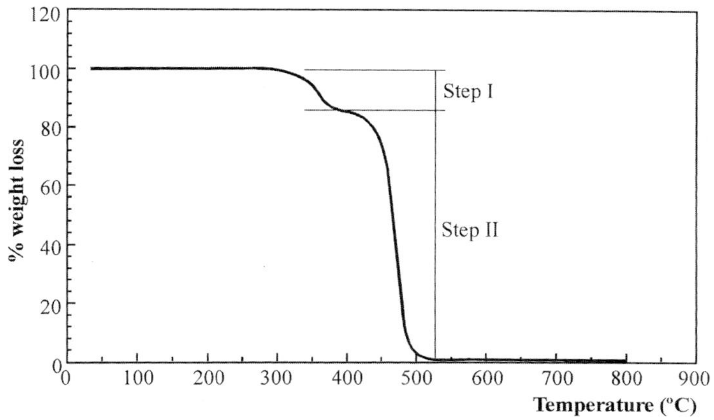

Figure 1. Thermogravimetric curve of EVA ALCUDIA ® PA-420. [I] Removing of volatiles generated in the degradation of the vinyl acetate, (mainly acetic acid). [II] Degradation of carbon chain residue.

The different DSC curves allow to identify the most suitable temperatures for processing by means of a hot plate press without EVA degradation (Figure 2). Since EVA degradation processes start at relatively low temperature (around 220 °C), it is necessary to process the mixtures at lower temperatures.

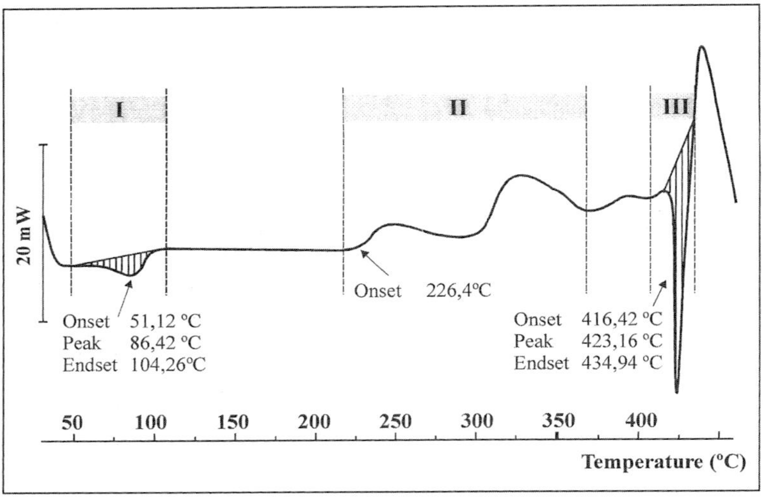

Figure 2. DSC thermogram of EVA with 50%wt Zn. [I] Melting of the crystalline phase of EVA, [II] Degradation process of EVA, [III] Melting point of Zn.

Mechanical properties of EVA/Zn system vary considerably with Zn content[3]. We observe an important reduction in mechanical ductile properties (Figure 3) such as elongation at break since Zn powder can act as a stress concentrator which plays an important role during the fracture process. This situation must be considered in order to use these composites at industrial level where a good balance between ductile and resistant properties is needed.

Similar behavior can be found in the evolution of the tensile strength (Figure 4) since the presence of Zn promotes the fracture due to the low interaction with the EVA matrix. This low interaction is related to the presence of polar groups in EVA chains which can interact with metallic Zn powder. Nevertheless, these interactions are very weak and do not contribute to improve adhesion between the two phases.

This lack of interaction in the interphase is responsible of the fracture process since there is not a continuous phase to transmit loads. Zn filler, which is quite resistant and promotes important changes in electrical properties, reduces considerably the mechanical ductile properties.

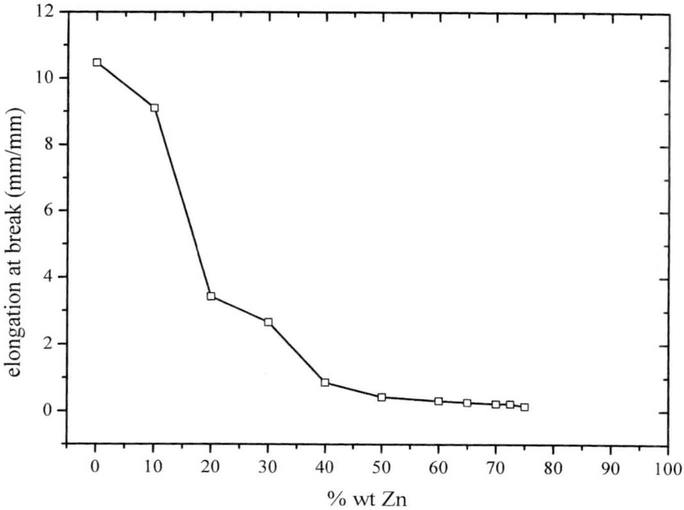

Figure 3. Variation of elongation at break of EVA-Zn system with different percentage of Zn.

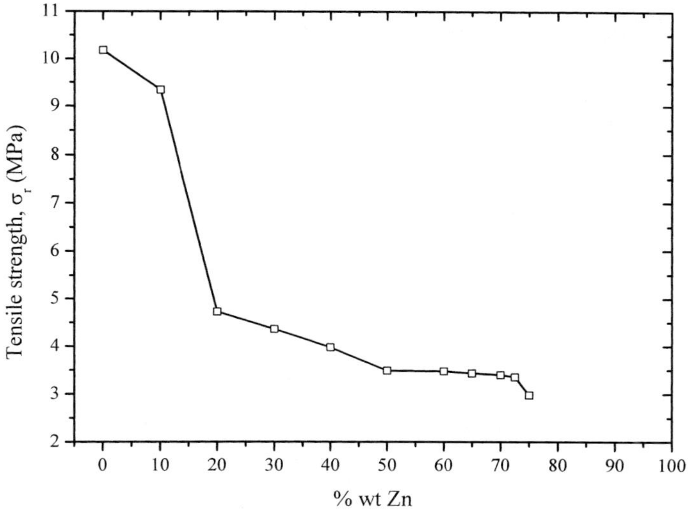

Figure 4. Variation of tensile strength of EVA-Zn system with different percentage of Zn.

Nevertheless, stiffness is increased by increasing the Zn loading in the mixture as indicated by the variation of Young modulus (calculated as tangent modulus) of the EVA-Zn system (Figure 5). Typical values of Young modulus of EVA copolymers are considerably increased to values similar to other commodities.

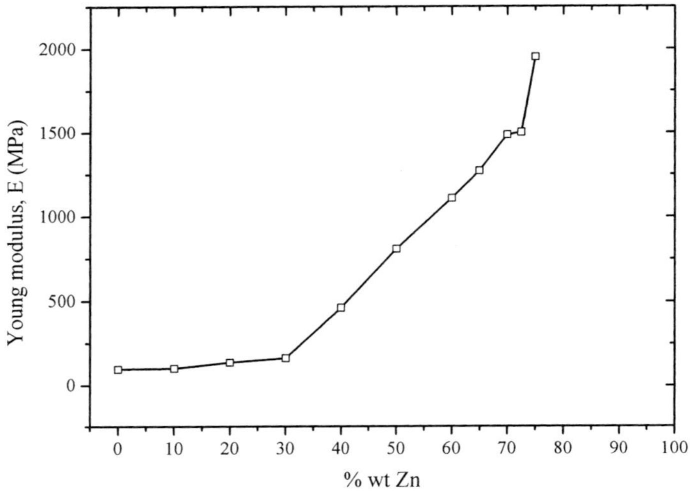

Figure 5. Variation of Young modulus of EVA-Zn system different percentage of Zn.

Prediction of behaviour

Due to the importance that fillers are acquiring in the preparation of composites, it is interesting to predict their behavior in order to optimize the formulations.[4-5]

There are many theories that can help us to establish the mechanical performance of the composites. Some of these theories are similar to those used to determine the performance of materials reinforced with different types of fibres. Many of these models consider different fractions of the material working in series or in parallel. Parallel fractions contribute to composite stiffnes while series fractions play an important rol in the fracture process since adhesion between phases must be considered.[6-7]

Percolation theory, which is of general use and applicable to all kind of multiphase systems, is one of the most interesting to study for prediction of the behavior of composites. When using this theory it is important to determine the type of net that better fits the behavior of the material (Figure 6) (square, triangular, hexagonal, cubic,...) since each one presents different percolation thresholds, necessary to predict the stiffness of the system.

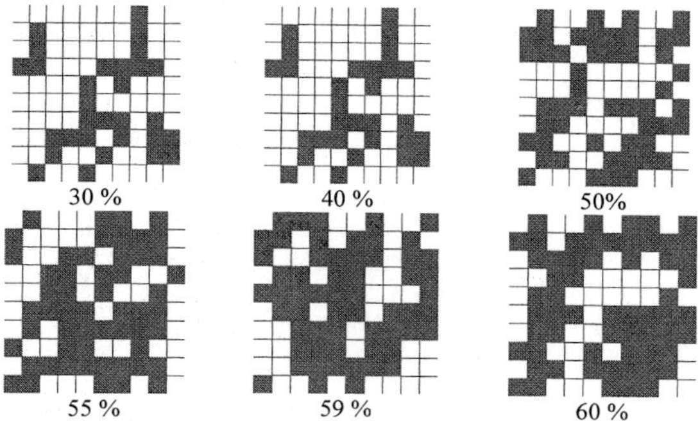

Figure 6. Percolation simulation for a square net (percolation threshold =0,59).

Percolation threshold based on mechanical properties is close to 0,16. This value can be obtained from the observation of the variation of the characteristics of the system. Electrical properties give similar results[8-9]. This value of threshold corresponds to spherical 3D systems and can be used to predict the mechanical performance, specially the stiffness, by means of the following expression:

$$E = E_0 \bullet (v - v_c)^T$$

where E_0 is a constant of the system, v is the fraction in volume of the major component, v_c is the percolation threshold, and T is the universal critical exponent. The values of percolation that give good results are 0,16 and 1,9 for v_c and T respectively pertaining to discrete domains of spherical geometry.

Conclusions

Composites based on EVA matrix with metallic Zn powder as filler are quite interesting materials for its use as electrodes due to their excellent electrical properties as described in previous works[9].

At the same time, the mechanical benefits of the EVA-Zn system are interesting and can be predicted by means of the use of general theories such as percolation theory, most useful to establish thresholds which determine the behavior of the different components of the system. Processing of these composites is quite easy, at low temperatures, due to the nature of both components. EVA must be subjected to a pulverization process to ensure good mixing with Zn powder.

[1] Y. Li, Y. Zhang, Y.X. Zhang, *Polymer Testing* **2004**, 23 (1), 83.
[2] Y.H. Hou, M.Q Zhang, M.Z. Rong et al., *Journal of Applied Polymer Science* **2002**, 84 (14), 2768.
[3] N.C. Das, T.K. Chaki, D. Khastgir, et al. *Polymer & Polymer Composites* **2000**, 8 (6), 395.
[4] J. Navarro et col., *Journal of Materials Science* **1994**, 29, 4604.
[5] J. Navarro, J. Vilaplana et col., *Journal of Electroanalytical Chemistry* **2000**, 33, 484.
[6] J. Kolarik and F. Lednicky, *Polymer Engineering and Science*, **1997**, 37.
[7] D. Stauffer & A. Aharony. "*Introduction to Percolation Theory*". Taylor & Francis Ltd (1994).
[8] Q.H. Zhang, D.J. Chen, *Journal of Materials Science* **2004**, 39(5), 1751.
[9] J.R. Subiela, J. López, et al, *Journal of Materials Science*, **2004** (In Press).

Macromol. Symp. **2005**, *221*, 217-225

Multiple Threshold Percolation in Polymer/Filler Composites

Douglas H. McQueen,*[1] *Karl-Michael Jäger*,[2] *Michaela Paligová*[3]

[1]Department of Materials Science and Engineering, Chalmers University of Technology, S-412 96 Göteborg, Sweden
E-mail: douglas@minmail.net
[2]Marketing and Development Centre Wire and Cable, Borealis AB, S-444 86 Stenungsund, Sweden
E-mail: Karl-Michael.Jager@borealisgroup.com
[3]Polymer Centre, Faculty of Technology, Tomas Bata University of Zlin, TGM 275 Cz-762 72 Zlin, Czech Republic
E-mail: paligova@ft.utb.cz

Summary: Local variations in filler particle concentration and/or shape and orientation in static filler/polymer composites are modelled as distributions of percolation thresholds. The concentration variations can be due to insufficient mixing, formation of semicrystalline voids during cooling from the melt, shrinkage during polymer curing, flow during physical compression or the like. Irregular filler shapes, especially elongated shapes, reduce the percolation threshold; thus natural variations in the shapes and orientations of filler particle aggregates lead to locally varying percolation thresholds. A distribution of percolation thresholds leads to an apparent percolation threshold based on the conductivity below the mean percolation threshold. For filler concentrations above the apparent percolation threshold the dielectric constant continues to increase before reaching a lowered peak value at the mean percolation threshold and then decreasing. Own experimental results on EBA /carbon black composites support the theory.

Keywords: carbon black; dielectric constant; percolation; polymer composite

Introduction

When electrically conductive filler particles are added to an insulating polymer matrix at first the conductivity of the composite rises only slowly. When the concentration of the filler particles is high enough they form a continuous but tortuous network across the whole composite, leading to a dramatic increase of the conductivity beginning from the percolation threshold. Also the dielectric permittivity of the composite increases up to the percolation threshold, where it peaks and then decreases in exactly the same way as it increased. This behaviour in both two and three dimensions is well understood and modelled in percolation theory (Efros and Shklovskii[1] Bergman and Imry[2], Stauffer[3, 4]) as confirmed in numerous experiments.

DOI: 10.1002/masy.200550322

However, in some less than ideal solid systems experimental results show that the permittivity continues to increase even above the percolation threshold. This has been observed by several investigators, among them McLachlan and Heaney[5] for carbon black in PE (polyethylene), McLachlan et al.[6] for powders on insulating grains, McLachlan et al.[7] for colloidal graphite coated polystyrene balls, Tchmutin et al.[8] for carbon black in epoxy resin, Jäger et al.[9] for carbon black in EBA (ethylene butylacrylate copolymer), Jäger, McQueen and Vilčáková [10] for carbon black in PMMA (polymethyl methacrylate) and Flandin et al.[11] for carbon black in epoxy. Characteristic of these static systems is that the polymers have either been cooled through a structural transition, allowed to cure after mixing the filler and polymer or physically compressed.

The question arises whether there is a problem with the experimental technique (for instance with the sample preparation) or with the theory (is percolation theory correct or is something lacking in the model?). Much effort has been expended in checking on and extending percolation theories and their associated scaling functions (Sarychev and Brouers[12], Balberg[13], McLachlan et al.[6], McLachlan et al.[7], Balberg[14]). Of especial interest is an exploration of the effects of allowing variations in the electrical characteristics of the gaps between conducting filler particles (Sarychev and Brouers [12], Balberg [13] 1998, Rubin et al. [15], Balberg [14], Jäger et al. [9], Jäger et al.[10]). To date no completely satisfactory quantitative explanation for "anomalous" results concerning the dielectric constant above the percolation threshold has been worked out in detail. Thus there may be a problem with the experimental technique (mixing).

As these experimental results are in fundamental disagreement with theory, it is important to sort out this problem.

In the following we show that a viable explanation of the discrepancy between theory and experiment is that the dielectric properties of the finite size solid composites are not homogeneous, but rather vary locally. This can be due to insufficient mixing, formation of semi-crystalline volumes during cooling from the melt, shrinkage during polymer curing, flow during physical compression, variations in the shape and orientation of the filler particles and their aggregates, etc. This is modelled as a distribution of percolation thresholds such that the dielectric properties of the whole finite composite sample are the sum of the properties of many small volumes in parallel with percolation thresholds distributed around an average percolation threshold. The results of the model provide an explanation of the experimental results.

Theory

The individual percolation curves were calculated using the most common formulae for the conductivity and the dielectric permittivity. The exact nature of the distribution (shape, width) of the percolation threshold distribution is not important for the purposes of the arguments put forward here. It is reasonable to assume a smooth percolation threshold distribution over a range of concentrations. Here a Gaussian distribution has been chosen with mean at 10.5% vol % and standard deviation 3.5 vol %. Twenty percolation curves as shown in Fig. 1 ("single percolation") spaced at filler concentration intervals of 0.7% were summed with the Gaussian weighting to give the left hand curves in Fig. 1 ("multiple percolation"). The lowest percolation threshold was then 3.5% and the highest one was 17.5%.

It is clear from the left hand curve for the conductivity in Fig. 1a that the apparent percolation threshold is about 5% rather than at 10.5% filler volume concentration. This shift from the "average" percolation threshold is related to the fact that the conductivity curve is asymmetrical around the percolation threshold. On the other hand, the left hand curve for the permittivity in Fig. 1b is symmetrical around 10.5%. The height of the permittivity against concentration curve decreases as the width of the percolation threshold distribution increases.

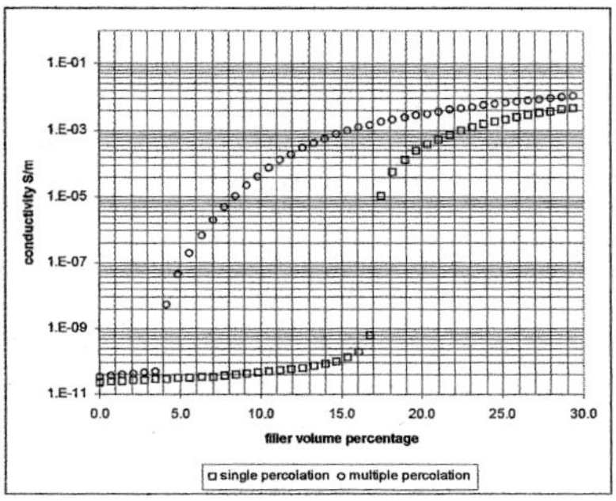

a)

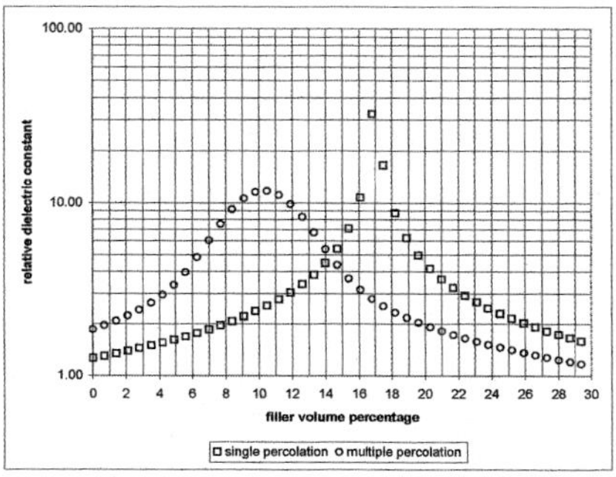

b)

Figure 1. The filler concentration dependence of a) the electrical conductivity and b) the relative dielectric constant for the case of (i) a single percolation threshold at 17% and (ii) multiple percolation thresholds between 3.5% and 17.5% by volume.

Figure 1 makes it clear that this distribution of percolation thresholds corresponding to parallel samples can produce results where the permittivity continues to increase at filler concentrations far above the apparent percolation threshold. The apparent critical exponent increases roughly according to t = 1.9 + 0.75 (std dev) with "std dev" the standard deviation of the Gaussian distribution in volume percentage. Below the percolation threshold the concentration dependence changes from a power law to exponential [12].

The above discussion has been concerned with filler concentration dependencies. For practical reasons it would be good to be able to extract information on the homogeneity of a composite from a single measurement on a sample. In principle such a procedure may be possible using frequency analyses.

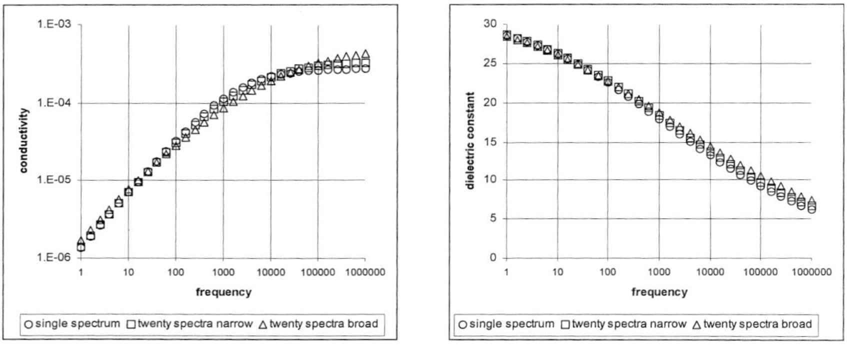

a) b)

Figure 2. The frequency dependencies of a) the conductivity and b) the relative dielectric constant for a single relaxation time, a narrow Gaussian distribution of relaxation times with a standard deviation corresponding to 100 and a broad Gaussian distribution of relaxation times with a standard deviation corresponding to 10^4 or four orders of magnitude.

For a given relaxation time τ the frequency dependencies of the conductivity and relative dielectric permittivities were modelled according to standard formulae. Figure 2 shows single characteristic relaxation time spectra compared to two multiple characteristic relaxation time spectra with Gaussian distributions of the logarithm of the relaxation times. The narrower distribution has a mean of - 4 (log 0.0001) with a standard deviation of 1 corresponding to a factor of ten and the broader distribution has the same mean and a standard deviation of 2 corresponding to a factor of one hundred. (Both distributions extend from minus two standard deviations to plus two standard deviations.) Thus in the latter case the variation in the characteristic relaxation time is over a factor of 10^4 from minus one standard deviation to plus one while the whole characteristic time spectrum starts from minus two standard deviations to plus two, corresponding to a factor of 10^8.

In spite of this broad variation in characteristic relaxation times especially the low frequency side of the spectra shown in Figure 2a for the conductivity are almost unchanged in shape. Only in the region of the transition to the high frequency plateau does the distribution have a significant effect. Figure 2b shows that the effect of the characteristic relaxation time distribution on the permittivity is also minimal. Essentially this is because even when there is a single characteristic relaxation time the spectrum already includes a broad distribution of Debye-type relaxation spectra, here covering something like four orders of magnitude. An important or perhaps irritating effect of the characteristic relaxation time distribution is to lower the slopes of the curves in Figures 2a and 2b, thus producing lower apparent critical exponents. This does not seem to be a good way of evaluation the homogeneity of the sample.

Experimental results for EBA and acetylene black

The pelletised thermoplastic samples consisted of an EBA matrix (similar to LDPE but with 4.3 mol% butylester side groups, density 925 kg/m^3) with varying amounts of acetylene black (carbon black) filler (density 1800 kg/m^3). They were press-moulded at 180 °C for thirty minutes. Then the two millimetre thick samples were cut to 15 millimetre diameter round discs and the flat sides covered with conductive paint. The dc conductivities of the samples were measured with a sensitive ohmmeter (Keithley 595 Quasistatic CV Meter). Finally, their capacitances and resistances were measured perpendicular to the flat faces as functions of the frequency with a Hewlet Packard 4284 A Precision LCR Meter in the range 20 Hz to 1 MHz.

Figure 3a shows the conductivity of the composites as a function of the filler concentration in weight percent (about double the concentration in volume percent). Each data point is the average of measurements on three samples. The apparent percolation threshold is between about 11 and 12 weight percent or about 6 vol %. The connected symbols are a theoretical fit to the data using a wide distribution of percolation thresholds (average percolation threshold 21 vol % corresponding to about 42 weight %) similar to that in figures 1 and 2.

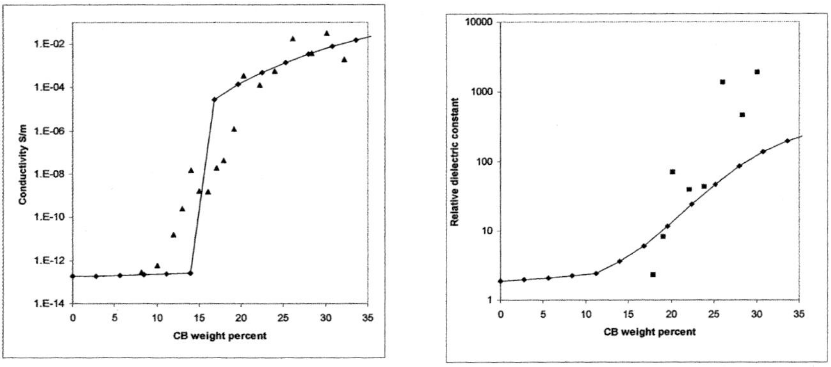

a) b)

Figure 3. Experimental results: a) the dc electrical conductivity versus the carbon black weight percent and b) the relative dielectric constant for 20 Hz versus the carbon black weight percent in thermoplastic EBA at room temperature as well as theoretical fits (connected symbols) to the data.

Figure 3b for the relative dielectric constant as a function of the filler concentration (again the average of three samples) shows that the dielectric constant continues to increase long after the apparent percolation threshold around 11 or 12 weight percent or about 6 vol % has been passed. The theoretical "fits" in figures 3a and 3b are only fair. A next step might be to vary the percolation threshold distribution to better fit the data. This potentially rather complicated task has not been attempted. A further step would be to relate the percolation threshold distribution to the filler particle distribution, as measured by some appropriate means.

Figure 4 show the frequency dependence of the conductivity with five representative spectra for filler concentrations from 8.14 weight percent to 30.16 weight percent. The slope of the low frequency part of the 8.14 weight percent curve is about 1.58, and the two spectra for the other samples have corresponding slopes of 1.95 and 1.64. As the concentration is significantly lower than the apparent percolation threshold, slopes approaching 2.0 are expected. The slope of the next curve (filler concentration 14.07 weight percent), still below the apparent percolation threshold, is about 0.88 (other sample spectra having slopes of 0.48 and 1.01). The following spectrum is for a concentration quite near the percolation threshold (17.07 weight percent) and the slope in the lower part of the curve is 0.44 (other curves having slopes of 0.34 and 0.81). The slope of the 22.17 weight percent curve is 0.45 (other curves having slopes of 0.42 and 0.53) in the high frequency segment. These slopes around 0.50 are a little lower than that predicted by the theory for single percolation thresholds (about 0.65) and are thus in better agreement with the multiple threshold slope prediction (about 0.55). Finally, the curve for 30.16 weight percent is almost flat, as expected.

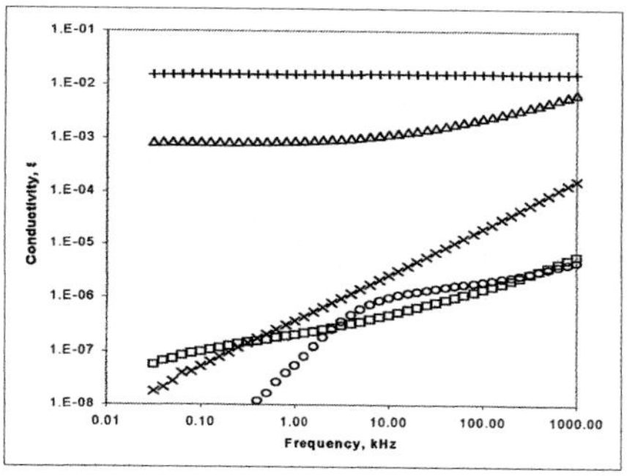

Figure 4. The frequency dependence of the electrical conductivity for different carbon black weight percentages in thermoplastic EBA at room temperature: (o) 8.14 w%, (x) 14.07 w%, (□) 17.07 w%, (Δ) 22.17 w% and (+) 30.16 w%.

These data show considerable scatter. Some of the imperfections may be due to measurement difficulties such as problems with measuring the dielectric permittivity of high conductivity samples, noise problems especially for samples with low conductivity, difficulty in achieving intimate electrical contact with the sample, etc. Other imperfections are due to real variations between samples that give rise to variations in the measurement results. This is intimately connected with the sample inhomogeneity and thus unavoidable.

Discussion and Conclusions

We propose that the static samples that show dielectric constants that increase with increasing concentration above the apparent percolation threshold are inhomogeneous on a scale of the thickness of the sample. The measurement results are interpreted as an average of the characteristics of a large number of small sample volumes measured in parallel. The different volumes have different filler concentrations, for instance due to formation of semicrystalline voids, stress fields or plastic flow during compression. When the filler particles are not spherical their shapes and orientation distributions reduce the percolation threshold. Thus irregularly shaped filler particles and filler particle aggregates lead to varying local percolation thresholds. This is often the case with intermediate structure or high structure carbon blacks. The ratio of the local filler concentration to the local percolation threshold can vary greatly on a laboratory scale. This leads to a concentration difference between the apparent percolation threshold and the peak in the concentration dependence of the dielectric constant, which can be at a much higher filler concentration.

The ideal case is a sample with the same statistical distribution of filler particles on all relevant scales throughout the whole sample. This is naturally very difficult to achieve in static (solid) samples, especially those in which the carbon black structures have a "high structure".[14, 15] Here we should like to emphasise that not only does the carbon black structure affect the percolation characteristics of the composites, but also the matrix polymer has a strong influence on the experimental results.[16] It is easier to achieve a "good" filler particle distribution in microemulsions[17, 18] or in two-dimensional samples prepared for instance by sputtering or etching.

A weakness of the present approach is that the distribution of percolation thresholds used to model the inhomogeneity of the composite can vary within wide limits as to shape and width. There are few physical restrictions on what can be possible. It is therefore difficult to test the model in detail. This does not mean that this sort of modelling is simply "sweeping the problem under the rug" since it is really possible, if not probable, that real samples often are inhomogeneous on the relevant scale. Outside the laboratory, in natural formations and structures, for instance rocks, one should surely expect that filler concentrations are variable within rather wide limits.

A disproof of the present multiple threshold model could be provided by a composite with homogeneous matrix and spherical, non-interacting filler particles randomly distributed throughout. If experimental results showing that the dielectric constant continues to increase significantly above a percolation threshold in the neighbourhood of 17% by volume were to be obtained, then the present multiple threshold model would be disproved. We know of no such example.

We conclude that a reasonable distribution of percolation thresholds in a sample can lead to a peak in the capacitance versus filler concentration curve at a higher concentration than that of the percolation threshold as determined from the conductivity versus filler concentration curve. The critical exponent t increases as the distribution of percolation thresholds broadens. Further, in the frequency plane the percolation threshold distribution can lead to a slight reduction of the apparent critical exponents x and y.

[1] A L Efros, B I Shklovskii, Physica Status Solidi b **1976**, *76*, 475
[2] D J Bergman, Y Imry, Phys. Rev. Let. **1977**, *39(19*), 1222
[3] D Stauffer, Phys. Reports **1979,** *54(1),* 1
[4] D Stauffer, "*Introduction to Percolation Theory*", Taylor and Francis, London 1985
[5] D S McLachlan, M B Heaney, Phys. Rev. B **1999**, *60*, 12746
[6] D S McLachlan, W D Heiss, C Chiteme, J Wu, Phys. Rev. B **1998**, *58*, 13 558
[7] D S McLachlan, Kefeng Cai, C Chiteme, W D Heiss, Physica B **2000**, *279*, 66
[8] I A Tchmutin, A T Ponomarenko, V G Shevchenko, N G Ryvkina, C Klason, D H McQueen, J. Polymer Sci. B: Polymer Phys., **1998**, *36*, 1847
[9] K-M Jäger, D H McQueen, I A Tchmutin, N G Ryvkina, M Klüppel, J. Phys. D: Appl. Phys. **2001**, *34*, 2699
[10] K-M Jäger, D H McQueen, J Vilčáková, J. Phys.D: Appl. Phys. **2002**, *35*, 1068
[11] L Flandin, T Prasse, R Schueker, K Schulte, W Bauhofer, J-Y Cavaille, Phys. Rev. B **1999**, *59(22)*, 14349
[12] A K Sarychev, F Brouers, Phys. Rev. Lett. **1994,** *73*, 2895
[13] I Balberg, Phys. Rev. B **1998**, *57*, 13 351
[14] I Balberg, Carbon **2002**, *40*, 139
[15] Z Rubin, S A Sunshine, M B Heaney, I Bloom, I Balberg, Phys. Rev. B **1999**, *59*, 12 196
[16] O Breuer, R Tchoudakov, M Narkis, A Siegmann, J. Appl. Polymer Sci. **1999**, *73*, 1655
[17] M T Clarkson, S I Smedley, Phys. Rev. A **1988**, *37*, 2070
[18] M T Clarkson, Phys. Rev. A **1988**, *37*, 2079

Macromol. Symp. **2005**, *221*, 227-236

Treatment of SMC Composite Waste for Recycling as Reinforcing Fillers in Thermoplastics

Didier Perrin, Eric Leroy, Laurent Clerc, Anne Bergeret, José-Marie Lopez-Cuesta*

Centre des Matériaux de Grande Diffusion (CMGD), Ecole des Mines d'Alès, 6 avenue de Clavières, 30319 Alès Cedex, France
Fax: (+33) 4 66 78 53 65; E-mail: eric.leroy@ema.fr

Summary: Needs for recycling thermoset composite materials are considerably increasing. Unfortunately, such materials are particularly difficult to recycle due to their crosslinked nature. In this paper, we focus on the case of Sheet Moulding Compounds (SMC), which consist of an unsaturated polyester filled with glass fibres and calcium carbonate. We consider the possibilities of mechanical treatments allowing to obtain ground fractions that could be used as fillers for the reinforcement of thermoplastics. Two processes are described and compared with an existing industrial one. Finally, optimised fractions of SMC waste are introduced into polypropylene, leading to promising results in term of mechanical properties.

Keywords: composites; fibre; grinding; recycling; SMC

Introduction

Thermosetting Polyester / Glass Fibre / Calcium carbonate composites, such as Sheet Moulding Compounds (SMC)[1], are high strength-to-weight ratio and low cost materials currently used in large series sectors such as automotive or electrical industry. Nevertheless, these composite materials are difficult to recycle[2]. This constitutes an important drawback for these two industrial sectors, which are subjected to recent and future European regulations asking for high material recyclability. At the same time, landfilling costs and regulations are becoming more and more incompatible with the current 10^6 tons per year European consumption of these materials. Consequently, there exists an urgent need for developing adapted recycling techniques[3,4].

The main obstacle to recyclability is coming from the crosslinked nature of the thermoset matrix, which makes its melting impossible, so that conventional technologies used for thermoplastics are not applicable. In the particular case of SMC composites, which composition consisted of 25 wt% of crosslinked polyester resin, reinforced by 25 wt% glass fibres and filled with 50 wt%

 DOI: 10.1002/masy.200550323

calcium carbonate, various recycling or valorisation strategies has been proposed : one way for SMC recycling is nowadays incineration with energy recovery, in which the organic matrix is burned, the inorganic residue being either landfilled or used as low grade aggregates. Technical investigations for incineration on this time are the more economically viable recycling solution. Chemical solvolysis of the matrix[5], or its sublimation in a fluidised bed process,[2,3] seem to become more promising techniques but they are still at the research stage. Mechanical recycling seems to be at this moment the only strategy that has reached industrial applications[4,6]. This technique is based on granulation and comminution of the composite material, possibly followed by a sieving process. The obtained fractions (fibres, powder) could be used for a variety of applications, depending on the particle size and glass fibre content : as fine powders (below 300 – 400 μm) could be for example used as fillers in thermoset compounds such as Bulk Moulding Compound (BMC), and long fibres (above 10 mm) in civil engineering materials[7,9-10], medium size fibres (about 1 mm) are recyclable in thermoplastics[11,12].

The present work is focused on this last promising[6-8] application, with the aim to develop a process allowing to obtain ground SMC fractions with optimised characteristics for their introduction as reinforcing fillers in thermoplastics. Two different treatment strategies, starting from SMC production waste are described and compared : the first strategy consists of a pure mechanical grinding where the compound is reduced in powder fractions, whereas the second strategy is a mixed process in which shredding, is followed by a partial dissolution of calcium carbonate using an acid solution. The results of these two approaches are compared in terms of process efficiency, glass fibre content of the obtained SMC ground fraction and glass fibre size distribution. These two last characteristics are of particular importance for the reinforcement of thermoplastics. Finally, the optimised selected fraction will be introduced in polypropylene (PP). The mechanical properties of the resulting composite will be compared to those of a commercial recycled SMC / PP composite and of a unrecycled glass fibres / PP composite.

Experimental

Materials :

500 × 500 × 4 mm^3 SMC composite panels provided by Saint Gobain Vetrotex International company (Chambéry, France) were moulded by Compositec SA (Le Bourget du Lac, France).

The resulting samples constitute model SMC production waste. Commercial short glass fibres (unrecycled fibres of average length 4,5 mm and diameter 15 μm), with a surface treatment compatible for an introduction into polypropylene matrix (PP) were also supplied by Saint Gobain Vetrotex International. Post-consumer commercial SMC ground fraction (T401) optimised for recycling as thermoplastic filler/reinforcement, was supplied by MCR (Tournon, France) as grey aggregates (about 0.5 and 1 cm in diameter) and used as reference.

Thermoplastic matrix used for composite manufacturing was supplied by Appryl under the trade name 3150 MN5, and consisted in mixture of polypropylene homopolymer (84 wt %) and (propylene-co-ethylene) 50/50 wt % copolymer (16 wt %).

Preparation of SMC ground fractions

The model SMC production waste sheets were first manually cut and conditioned into smaller parts (50 x 50 x 4 mm^3), which were then treated by one of the two processes described on Figure 1, leading to two different filler fractions : SMC^{mod}_1 and SMC^{mod}_2.

The filler SMC^{mod}_1 was obtained from a pure mechanical grinding. The SMC pieces were ground using a Rotoplex-Alpine Rotary Cutter Mill working at a 1500 rpm and using a three steps procedure (Figure 1). Each step consisted in grinding the material during 25 min, the ground fraction passing through a grid being ground again in the next step. Three different grids were used, with progressively reduced hole diameters (12 mm for step 1, 6 mm for step 2, and finally 4 mm for the step 3).

The filler SMC^{mod}_2 was obtained from a mixed process consisting in two steps. The SMC pieces were first ground during 25 minutes at a 1500 rpm using the same Rotoplex-Alpine Rotary Cutter Mill, with a grid of 12 mm. The passing ground fraction was then introduced in a H_3PO_4 buffer solution with strong agitation in order to partially dissolve the calcium carbonate. After that, the filler was filtered and dried. The objective of this second treatment was to obtain the highest fibre fraction and fibre length in the final filler. Two experimental devices were used to optimise the acid concentration, the agitation time and the conditions of sieving. So, the following dissolution conditions were obtained : 100 g of shredded SMC was dissolved at room temperature into a H_3PO_4 buffer solution (0.3 M, pH 1.9, 5 l). The stirring time and the rotative speed of the propeller agitator were respectively of 60 min and 1980 rpm. The partially dissolved fraction was then filtered and dried for 3 h at 80°C and 7 h at 60°C.

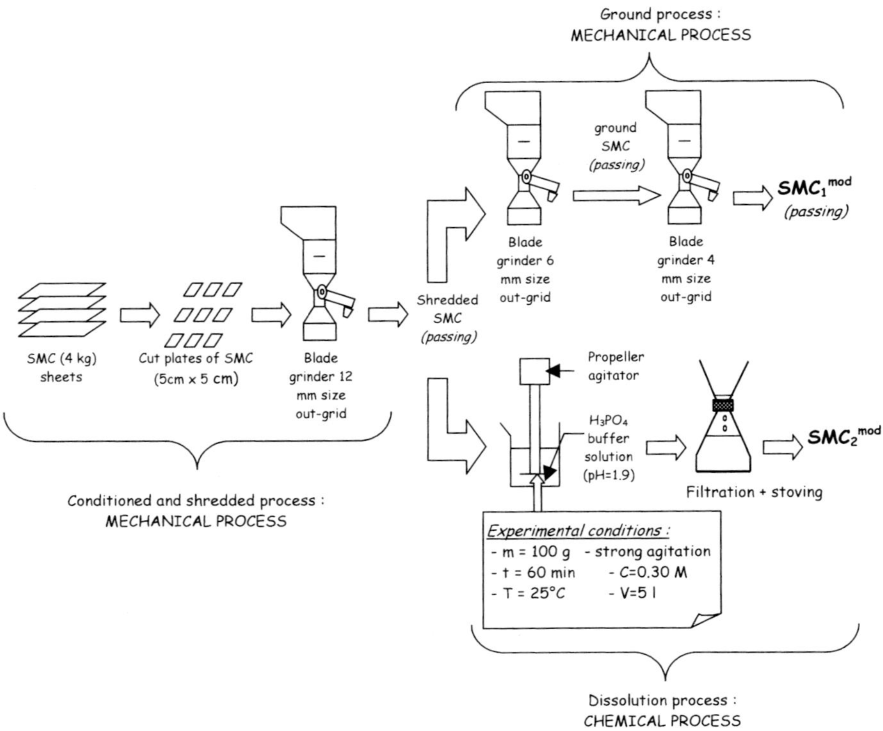

Figure 1. Preparation of fillers from SMC production.

Characterisation of the different recycled fractions (SMC^{mod}_1, SMC^{mod}_2, T401)

SMC^{mod}_1 , SMC^{mod}_2 and T401 samples were first observed by Scanning Electron Microscopy (JEOL JSM 35CF) in order to analyse the morphology of the samples. Then, the composition of these recycled fractions and of the initial SMC panels was determined as follows :

(i) The organic fraction, mainly composed of the polyester resin matrix, was first measured by weight loss after pyrolysis at 550°C for 180 min ;

(ii) $CaCO_3$ amount was then measured by calcimetry. This technique consisted in the dissolution of about 1- 2 g of SMC fraction in a HCl solution and the measurement of CO_2 volume evolved from the reaction with $CaCO_3$. For experiments which were investigated in standard temperature and pressure conditions, the amount of calcium carbonate present in the filler can be determined by the following equation (1) :

$$\%CaCO_3 = \frac{100.M_{CaCO_3}.V_{CO_2}}{24040.m_{SMC}} \qquad (1)$$

where m_{SMC} is the SMC amount, V_{CO2} the CO_2 volume evolved and M_{CaCO3} the molar mass of $CaCO_3$.

(iii) Finally, the remaining glass fibre fraction was calculated by mass balance. The process efficiency is determined by the following equation (2) :

$$R_m = \frac{m_p}{\sum_p m_p + \sum_r m_r + \sum_{per} m_{per}} \qquad (2)$$

where m_p is the mass of passing elements, m_r the mass of oversize elements and m_{per} the mass of the overall wastes.

Fibre length distribution was determined both before and after composite processing for all the recycled fractions ($SMC^{mod}{}_1$, $SMC^{mod}{}_2$ and T401) and for the commercial short glass fibres after extrusion. In all cases, the material was first pyrolised and put in contact with HCl in order to eliminate organic and calcium carbonate residues. About 1 mg of the remaining fibres was dispersed into water to a roasted glass of porosity 4 (hole diameter of 5 – 15 µm) which prevented the crossing of fibres through the filter with glass plates on the bottom. The system was then put under vacuum in order to evaporate the water, so that the fibres would cast onto the glass plates deposited into the roasted glass. This procedure allowed to obtain well separated fibres, without using any solvent as a difference with techniques described in litterature[8, 15].

Another advantage of this method is that no shear is applied on the fibres which are relatively brittle in comparison with another technique, in which fibres are put between two glass plates with polyethylene glycol and dispersed by shear[13].

For each sample, 10 glass plates with dispersed fibres were prepared. 30 photographs per sample observed through Wild Heerbrugg M8 binocular were taken with a high resolution Sony Iris CCD cine-camera to have a 800 fibres population[16 - 19]. Analogical signal to numeric one conversion was carried out by a IC-PCI numerisation card so that images size of 768x572 pixels were obtained. Fibres length distribution was then obtained by a semi-automatic images analysis system based on Optimas software. Number average length L_n, length average length L_l and dispersity were then calculated using equations (3 to 5).

$$L_n = \frac{\sum_i \frac{1}{P_i} L_i}{\sum_i \frac{1}{P_i}} \tag{3}$$

$$L_l = \frac{\sum_i \frac{1}{P_i} L_i^2}{\sum_i \frac{1}{P_i} L_i} \tag{4}$$

$$D = \frac{L_l - L_n}{L_n} \tag{5}$$

Composite processing :

The different PP composites studied are listed on Table 1. The filler contents were chosen in order to have the same final 20 wt % glass fibre content to the composite. It can be observed that it was necessary to introduce 47 wt % of SMC^{mod}_1 or SMC^{mod}_2 fractions in order to obtain a final filler content of 20 wt % so that it would be difficult to process composites systems with higher fibre contents. A twin screw extruder was used to compound the various fibrous samples into PP (Clextral BC 21, 200°C, 350 min^{-1}). The screw profile was optimised to reduce glass fibre breaking during the process with an adopted extrusion process (2 zones of light mixing and a long pass-through zone).

Table 1. Composition of the different composites in terms of fibre amount.

Composites based on PP	Commercial glass fibres (1)	Pure grinding process (SMC^{mod}_1) (2)	Mixed process (SMC^{mod}_2) (3)	T401 (MCR) (4)
Fillers amount (% wt)	20	47	47	40
Overall fibre amount (% wt)	20	20	20	20

(1) PP20GF (2) $PP20GFsmc^{mod}_1$ (3) $PP20GFsmc^{mod}_2$ (4) $PP20GF_{T401}$

The melted compounds were then chipped into pellet form after cooling. As in the extrusion step, the composites were dried[14] for 3 days at 50°C prior to injection moulding for allowing polymer chains relaxation. Standard ISO 527-2 type 1A tensile test specimens were moulded using a 95 t SANDRETTO AT injection system.

Mechanical properties of the composites :

Tensile and Charpy impact tests were performed 3 days after injection moulding process[14]. Tensile tests were carried on using a ZWICK Z010 apparatus according to ISO 527 Standards to measure Young's modulus and yield stress. Charpy tests were investigated on un-notched standard specimens using a ZWICK 5101 (7 Joules) setup according to ISO 179 Standards.

Results and discussion

Characteristics of fillers compounds :

The SEM observation performed on SMC^{mod}_1, SMC^{mod}_2 and T401 fractions showed that most of the glass fibres present have been well separated from the polyester matrix, in agreement with the results obtained by Allred et al.[11]. Nevertheless, as Kouparitsas et al.[12], can be still observed some aggregates containing polyester and fibres arranged in beams (Figures 2a). Calcium carbonate particles seemed to be present on the fibre surface (Figures 2b) which could provide anchorage sites to polypropylene chains.

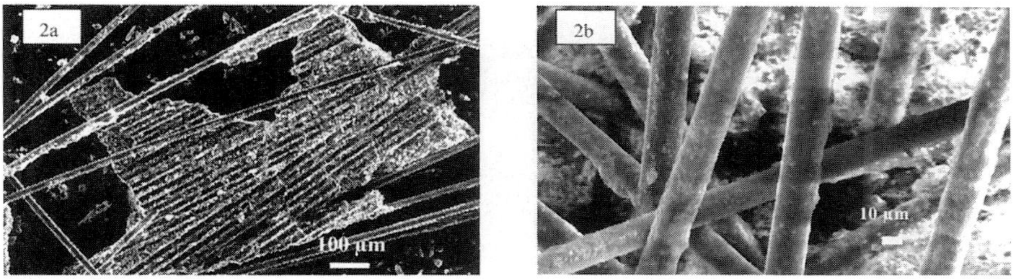

Figure 2. SEM images x100 (2a) and x500 (2b) of passing through ground SMC^{mod}_2.

The compositions of initial SMC states and of SMC^{mod}_1 , SMC^{mod}_2 and T401 recycled fractions are presented on Table 2. It can be observed that the use of a mixed process increases the glass fibre content of about 47 % compared to the pure grinding process. The high glass fibre content observed for T401 (42 %) is ascribed to the fact that this fraction is a mixture of pure glass fibre waste and SMC recycled fractions and that a sieving step was added in the process compared to our pure grinding process. Moreover, this additional step could have a negative effect on the global efficiency of the process which is about 60% compared to our processes that lead to higher efficiencies (70 and 90 %). SMC^{mod}_2 specimen seemed to show a greater amount of long fibres and smaller amounts of polyester and calcium carbonate particles. No significant differences are given in evidence between T401 and SMC^{mod}_1 fractions.

Table 2. SMC composition and efficiency according to the recycling process.

Composition (weight %)	Initial SMC	Pure grinding (SMC^{mod}_1)	Mixed process (SMC^{mod}_2)	T401 (MCR)
Unsaturated Polyester	26.0 ± 2.5	22.7± 2.5	31.8 ± 2.5	22.0 ± 2.5
Calcium carbonate	52.0 ± 2.5	55.0 ± 2.5	25.6 ± 2.5	29.0 ± 2.5
Glass fibre	22.0 ± 2.5	22.3 ± 2.5	42.6 ± 2.5	42.0 ± 2.5
Efficiency (%)		90	70	60

The glass fibre length distribution characteristics are shown on Table 3. SMC^{mod}_2 clearly showed higher fibre length than SMC^{mod}_1 and T401. The fibre length is moreover too high to be analysed quantitatively by our image analysis procedure. Nevertheless it can be noticed this fibre length is comparable to that of the commercially available short glass fibres used for reinforcing thermoplastics. SMC^{mod}_1 has an average fibre length comparable to that of T401, which has a smaller dispersity ; this is also enabled to the sieving operation.

Table 3. Glass fibre distribution according to the recycling process.

Distribution of glass fibre	Initial SMC	Pure grinding (SMC^{mod}_1)	Mixed process (SMC^{mod}_2)	T401 (MCR)
Average length (mm)	25 - 50	0.62	2 - 10	0.69
Dispersity	-	0.61	-	0.29

Composites properties :

The glass fibre length distributions were determined after compounding by extrusion. Table 4 results showed that the final length distribution obtained with SMC^{mod}_2 was similar to that obtained with commercial glass fibres and much higher than with T401. These preliminary data may indicate that the degradation rate of our recycled fibres fractions is not higher than those of standard short glass fibres ; and these, despite of the fact that recycled fibres should no more brittle because of all the drastic conditions of grinding mechanical properties of the different composites are shown on Table 5. It can be observed that all mechanical properties (Young's modulus, yield stress and impact resistance) are decreased when recycling fibre fractions (SMC^{mod}_2 and T401) are used compared to commercial glass fibres.

Table 4. Glass fibre length distribution after incorporation in PP matrix by extrusion process.

Distribution of glass fibre in PP	Commercial glass fibres	Mixed process ($SMC^{mod}{}_2$)	T401 (MCR)
Average length in number (mm)	1.20	1.20	0.20
Average length in length (mm)	1.50	1.50	0.25
Dispersity	0.21	0.21	0.29
Critical length of glass fibres (mm) (according to Cox's model)[20]	0.60	0.60	0.60

This result may be due to the presence of residual crosslinked polyester particles (about 14.9 wt% for $PP20GFsmc^{mod}{}_2$ and 8.8 wt % for $PP20GF_{T401}$) and of $CaCO_3$ (about 12.0 wt % for $PP20GFsmc^{mod}{}_2$ and 11.6 wt % for $PP20GF_{T401}$) within the composite material. These residues may be considered as impurities without any compatibility towards polypropylene chains that may increase the probability to cracks formation. Moreover these residues may modify the glass fibres orientation during injection moulding because of different melt viscosities and shear stresses. Finally, the lowered mechanical properties may be due, on the one hand to the brittleness of recycled fibres because of grinding conditions compared to standard short glass fibres, and on the other hand, to the lack of adhesion promotion on the recycled fibres surface. Next investigations would be to quantify the intrinsic mechanical properties of a recycled fibre and to analyse the recycled fibres surface though surface analysis techniques such as XPS (X ray Photoelectron Spectroscopy) and chemical extraction procedures[21]. Our alternative for better mechanical properties with recycled fibres fractions would be to improve the adhesion between the polypropylene matrix and the recycled fibres by introducing compatibilisation agents or by thermal or physical or chemical surface treatment. Table 5 showed also higher tensile properties for $PP20GFsmc^{mod}{}_2$ than for $PP20GF_{T401}$ composite materials (as impact resistance kept constant with regards to standard deviations).

Table 5. Mechanical properties of different composite based PP matrix.

Mechanical properties	Virgin PP	PP20GF	$PP20GFsmc^{mod}{}_2$	$PP20GF_{T401}$
Tensile properties (Young's modulus) (MPa)	1414 ± 45	5076 ± 45	4175 ± 279	3922[8]
Yield stress (MPa)	27.7 ± 0.6	57.6 ± 0.4	30.2 ± 0.4	25.8[8]
Charpy resilience (non notched test pieces) (kJ/m^2)	don't break	18.4 ± 1.9	8.3 ± 0.8	10.3[8]

This result may be due to the very low fibre lengths measured after extrusion compounding for $PP20GF_{T401}$ (0.2 mm) compared to $PP20GFsmc^{mod}{}_2$ (1.2 mm). This low fibre length should be the consequence of more brittle fibres after MCR recycling process so that a classical composite processing broke them easily. Nevertheless a such important variation in fibres length has also a non negligible effect on fibres orientation and dispersion and on the interfacial area between polymer and fibres, so that it remains delicate to conclude at the present moment.

Conclusions

A new efficient treatment process for obtaining filler compounds from SMC production waste has been described in this paper. Compared to existing processes involving only mechanical grinding, the addition of a dissolution step for calcium carbonate in our process allowed to obtain high glass fibre content in the final filler compound (about 43 wt %) with a reasonable overall efficiency (70%). This final composition was comparable to that obtained by industrial processes. Nevertheless, our process lead to higher glass fibre length in the compound, which are similar to those of fibres commercially available for introduction into thermoplastics. Nevertheless, the introduction of the obtained filler compound into a polypropylene matrix lead to a decreased in tensile and impact properties. Promising results should be expected taking into account the fact that the filler-matrix interface adhesion has not been optimised yet. This point will be the subject of a future paper.

[1] A.G. Pedroso, *Progress in Rubber, Plastics and Recycling Technology*, **2002**, *18(2)*, 111.
[2] J.R. Kennerley, R.M. Kelly, *Composites Part A*, **1998**, *29A*, 839.
[3] S.J. Pickering, R.M. Kelly, *Composites Science and Technology*, **2000**, *60*, 509.
[4] Recycling takes off at MCR, *Composites International*, **2002**, *48*, 27.
[5] H. Winter, H.A.M. Mostert, *Journal of Applied Polymer Science*, **1995**, *57(11)*, 1409.
[6] C.E. Bream, P.R. Hornsby, *Journal of Materials Science*, **2001**, *36*, 2965.
[7] Thermosetting materials: recycling and applications, *Composites (Paris)*, **1999**, *35*, 49.
[8] J.N. Reygrobellet, *PhD thesis*, University of Montpellier, France, **2000**, 234 p.
[9] Recycling takes off at MCR, *Composites International*, **2002**, *48*, 27.
[10] C. Moncel, *Recyclage Récupération Magazine*, **2002**, *16*, 19.
[11] R.E. Allred, L.D. Busselle, *Journal of Thermoplastic Composite Materials*, **2000**, *13*, 92.
[12] C.E. Kouparitsas, C.N. Kartalis, *Polymer Composites*, **2002**, *23(4)*, 682.
[13] L. Averous, *PhD thesis*, University of Montpellier, France, **1995**, 193 p.
[14] D.A. Steenkamer, J.L. Sullivan, *Composite Part B*, **1998**, *29B*, 745.
[15] R. von Turkovitch, L. Erwin, *Polymer Engineering and Science*, **1983**, *23(13)*, 743.
[16] M.J. Carling, J.G. Williams, *Polymer Composites*, 1990, 111(6), 303.
[17] F. Ulrych, M. Sova, *Polymer Composites*, **1993**, *14(3)*, 229.
[18] V.B. Gupta, R.K. Mittal, *Polymer Composites*,**1989**, *10(1)*, 8.
[19] V.B. Gupta, R.K. Mittal, *Polymer Composites*, **1989**, *10(1)*, 16.
[20] J.L. Thomason, *Composites Part A*, **2002**, *A33*, 1283.
[21] C.D. Papaspyrides, J. Poulakis, *Polymer for Advanced Technologies*, **1992**, *3*, 367.

Rubber Powder – A Perspective Filler of Thermoplastics

Udo Wagenknecht,[*1] Stefanie Steglich,[1] Sven Wiessner,[2] Hannes Michael[2]*

[1] Institute of Polymer Research Dresden, Hohe Str. 6, 01069 Dresden, Germany
Fax: (+49) 351 4658 290; E-mail: wagenknt@ipfdd.de
[2] Chemnitz University of Technology, Department of Plastics Processing, 09107 Chemnitz, Germany
Fax: (+49) 371 531-3776 ; E-mail: sven.wiessner@mb.tu-chemnitz.de

Summary: Rubber powder, ground at ambient temperature, was compounded with polypropylene using only 16 L/D of a co-rotating twin screw extruder. An optimum of extruder configuration and selected process parameters were found for compounds with a rubber filling rate of 50 wt.-%. Restrictions of a reactive compounding were taken into account. The dispersion of the rubber powder is not a limiting process. A compound containing 70 wt.-% rubber powder showed to be overfilled.

Keywords: compounding; particle size distribution; polypropylene; recycling; rubber

Introduction

The amount of used tyres is 600 000 tons/a only in Germany. They are mostly burnt in energetic recycling processes e.g. for cement production. A higher level of environmental protection would be an use as material. Modern techniques give the possibility to grind the rubber into particles in micrometer-scale. Such rubber powder can be an interesting filler for plastics in account to its elasticity.

An only physical compounding of rubber powder with different thermoplastics results in products with a low or intermediate mechanical behaviour. This is due to the low interaction between the surface of the cross-linked rubber particles and the polymer matrix resulting in low adhesion. A chemical bonding between the surface particles and the matrix is necessary for a substantial improvement in mechanical properties. As the rubber surface particles contain residual unsaturated double carbon-carbon bonds, a grafting of the matrix polymer onto the rubber surface initiated by radical donators, e.g. peroxides, could improve the rubber-polymer interaction.

 DOI: 10.1002/masy.200550324

A process of reactive compounding of rubber powder with polypropylene in an internal mixer has been developed[1,2,3]. A variety of organic alkyl- and arylperoxides, e.g. di(tert-butyl) peroxide (DTBP) or dicumyl peroxide (DCUP), were used as radical donators for the development of the discontinuous reactive compounding process in the internal mixer. It is well known that polypropylene undergoes β-scission in the presence of radicals, in contrast to polyethylene, which undergoes a cross-linking reaction. Therefore, the additional use of the so called radical catchers or radical acceptors, such as phenolics or molecular sulphur, could prevent the unwanted scission of the polypropylene backbone. A good level of mechanical properties of the reactive compounds could also be achieved even without adding radical catchers, if the rubber particles act as free radical scavengers.

Figures 1 and 2 show the difference in the tensile strength and elongation at break obtained from stress-strain tests according to ISO 527 - standard for physical and reactive compounds (so called elastomeric alloys).

The reaction time is one of the main process parameters in case of reactive compounding. The control of the reaction time is quite easy by using an internal mixer. However, the processing costs using an internal mixer are high which leads to a low competitiveness caused by price of the products. That's why the final aim of this work is the design of a continuous compounding process for reactive polypropylene- rubber powder compounds.

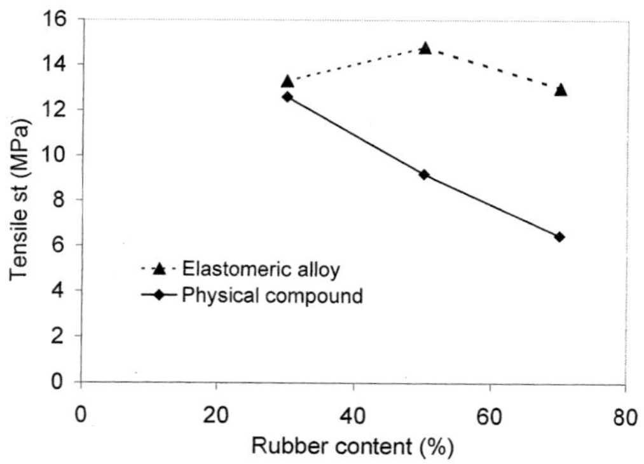

Figure 1. Tensile strength vs. rubber content of physical and reactive compounds.

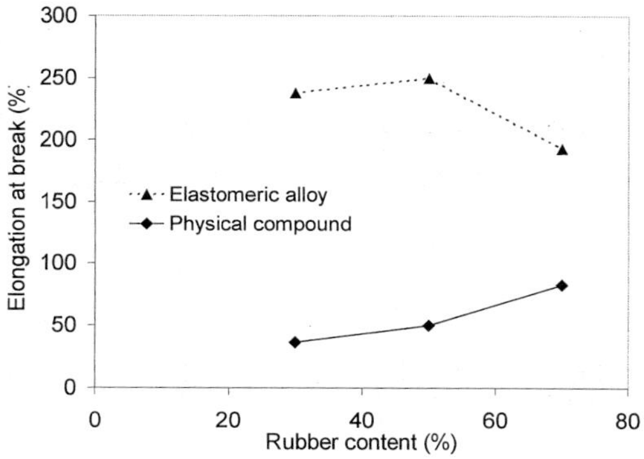

Figure 2. Elongation at break vs. rubber content of physical compounds and reactive compounds.

Experimental

Two main problems have to be solved for the design of a continuous reactive compounding process of PP with rubber powder:

- Dispersion and distribution of rubber powder in a PP-melt
- Reaction of PP with rubber powder, set off by a peroxide activator

The experiments described below are related to the powder dispersion and distribution in a twin screw extruder.

Figure 3. SEM micrograph of the rubber powder.

A rubber powder obtained from tread layers truck tyres by grinding at ambient temperature was used. The structure of the powder is shown in the SEM-micrograph of Figure 3 and the particle size distribution obtained by laser diffraction is given in Figure 4. A polypropylene-ethylene-copolymer with a melt flow rate of 8 g/10 min (190 °C/2.16 kg) and a comparatively low melting point of 139 °C was used as the thermoplastic matrix.

For the experiments a Micro 27 made by Leistritz, Nuremberg in Germany, was used as the compounding unit. This unit is a tightly intermeshing co-rotating twin screw extruder with a screw diameter of 27 mm and a variable L/D ratio from 24 up to 44. The long variant of L/D 44 will be used for the complete reactive compounding process. The L/D ratio of 24 was used in this work for the investigation of dispersion and distribution of rubber powder without the reactive processing step.

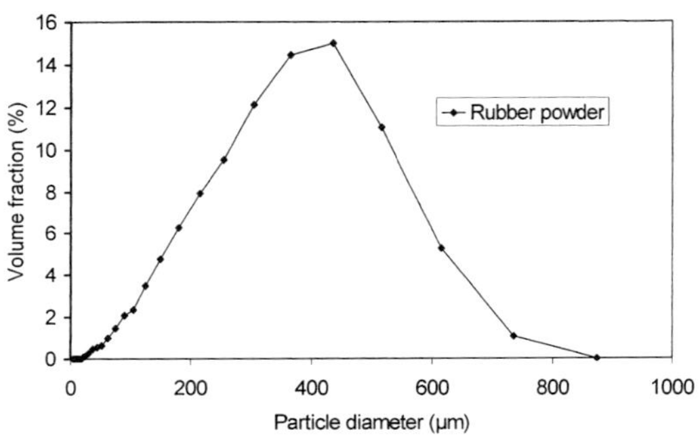

Figure 4. Particle size distribution of the rubber powder.

Both components were fed separately by gravimetric dosage units into the hopper. The fourth barrel element of the extruder (approximately 14 L/D) was closed with a closure part. This barrel element could be opened to take samples of the PP-rubber powder-melt at this place. After the process parameters were balanced, the extruder was stopped for taking samples.

The process was varied in relation to:

- Screw design (length of the kneading zone 60 and 30 mm, Figures 5 and 6)
- Screw speed (200 and 100 rpm)
- Temperature profile of the barrel T_{bmax} (2 different profiles)
- Feed rate (6 and 12 kg/h)
- Rubber powder content (50 and 70 wt.-%)

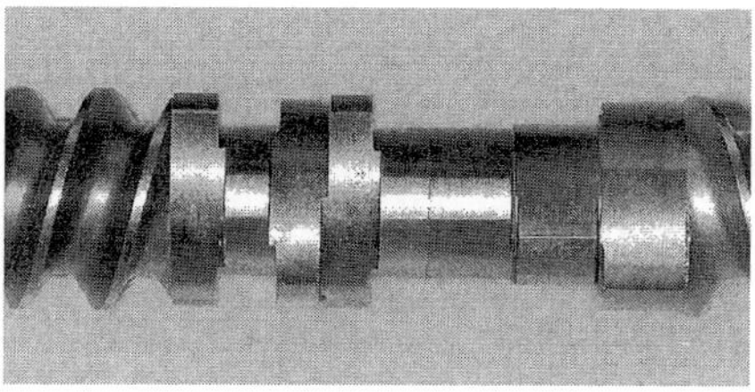

Figure 5. Kneading zone for intense kneading.

Figure 6. Kneading zone for medium kneading.

After taken the samples out of the extruder they were pressed to sheets of approximately 5 mm thick. Small pieces of the sheet were cut from one side with a cryo-microtom. The flat surfaces were partially imaged using a low voltage scanning microscope DSM 982 GEMINI Carl Zeiss. The low voltage option of this microscope made it possible to work with untreated samples. A thin gold film on the surface, usually necessary for polymers at standard voltages, leads to a very poor contrast between rubber particles and matrix polymer.

The micrographs were analysed computer aided using a special image analysing software (Analysis). Five different micrographs were used for every sample. The particle size distribution of the rubber powder before compounding was obtained using a laser diffraction sizer (Sympatec).

Results and Discussion

Four main parameters were selected to evaluate the different process parameters according to the final aim – design of a reactive compounding process:

- **Feeding/conveying behaviour:**

  The feeding of the voluminous rubber powder together with the polymer pellets in the main hopper can result in conveying problems. This variant of feeding was chosen to keep the extruder length which is necessary for physical incorporating of the rubber powder as short as possible.

- **Mass temperature**

  A small temperature window is caused by the melt temperature of polypropylene and the limiting temperature of the activator system for the reactive compounding.

- **Torque reserve**

  The reactive compounding results in an increase of melt viscosity. Therefore, the physical compounding should consume only a part of the maximum torque.

- **Dispersion of rubber powder**

  The evaluation of dispersion effectiveness using the particle size distribution in the compounds made no sense. All particle size distribution curves of the compounds were very similar, as shown in Figure 7. It was observed that none of the compounds exhibit a completely dispersed morphology, i.e. each compound exhibits a certain amount of contact points between the rubber particles. Contacts between particles will be weak zones in the final products with respect to the mechanical properties. That's why the number of observed contacts per area was used for the evaluation of the dispersion success. Such contacts can be observed in Figures 8 and 9.

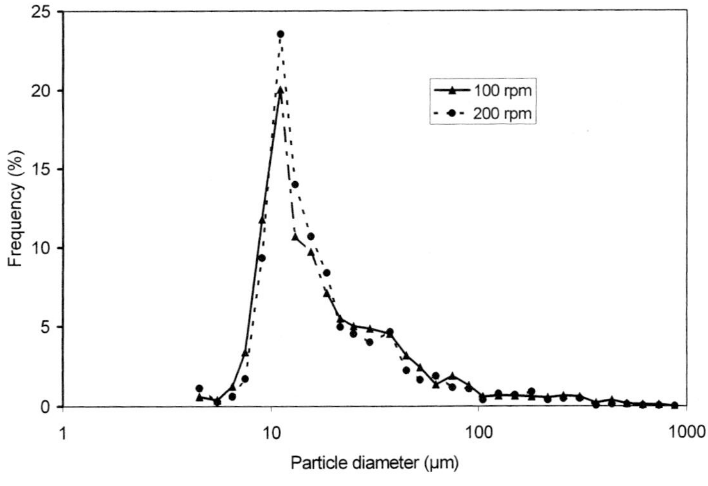

Figure 7. Examples of particle size distribution after compounding at different screw speeds (filling rate 50 wt-%, medium kneading, low barrel temperature).

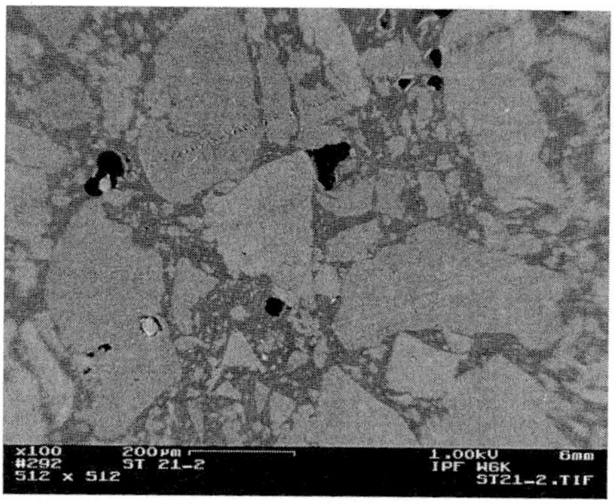

Figure 8. SEM micrograph of a compound containing 70 wt.-% of rubber powder.

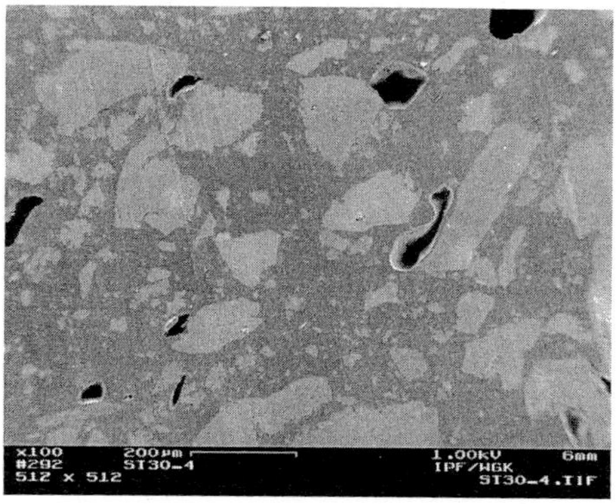

Figure 9. SEM micrograph of a compound containing 50 wt.-% of rubber powder.

Feeding/conveying behaviour (FC)	easy flow	unstable flow	tailback
Temperature window (activator system) (TW)	160 – 165 °C	166 – 170 °C	> 170°C
Torque reserve for reactive compounding (TR)	> 60%	50 – 60%	< 60%
Dispersion of rubber powder (DS)	< 20 contact points/cm²	> 20 contact points/cm²	Agglomerates > 10 particles.

Figure 10. Classification of results.

For a better illustration of the influence of the four main parameters on the extrusion process and product quality, the results were collected into three classes as shown in Figure 10. Light grey indicates a stable process, medium grey indicates probable problems for the reactive compounding, and dark grey indicates parameter settings impossible for the reactive compounding.

		Intense kneading		Medium kneading	
		n=100 rpm	n=200 rpm	n=100 rpm	n=200 rpm
$\dot{m}$ = 6 kg / h	T_{bmax}=160°C	FC	FC	FC	FC
		TW	TW	TW	TW
		TR	TR	TR	TR
		DS	DS	DS	DS
	T_{bmax}=150°C	FC	FC	FC	FC
		TW	TW	TW	TW
		TR	TR	TR	TR
		DS	DS	DS	DS
$\dot{m}$ = 12 kg / h	T_{bmax}=160°C	FC	FC	FC	FC
		TW	TW	TW	TW
		TR	TR	TR	TR
		DS	DS	DS	DS
	T_{bmax}=150°C	FC	FC	FC	FC
		TW	TW	TW	TW
		TR	TR	TR	TR
		DS	DS	DS	DS

Figure 11. Illustration of possible processing windows for compounds containing 50 wt.-% rubber powder.

		Intense kneading		Medium kneading	
		n=100 rpm	n=200 rpm	n=100 rpm	n=200 rpm
$\dot{m}$ = 6 kg / h	T_{bmax}=160°C	FC	FC	FC	FC
		TW	TW	TW	TW
		TR	TR	TR	TR
		DS	DS	DS	DS
	T_{bmax}=150°C	FC	FC	FC	FC
		TW	TW	TW	TW
		TR	TR	TR	TR
		DS	DS	DS	DS
$\dot{m}$ = 12 kg / h	T_{bmax}=160°C	FC	FC	FC	FC
		TW	TW	TW	TW
		TR	TR	TR	TR
		DS	DS	DS	DS
	T_{bmax}=150°C	FC	FC	FC	FC
		TW	TW	TW	TW
		TR	TR	TR	TR
		DS	DS	DS	DS

Figure 12. Illustration of possible processing windows for compounds containing 70 wt.-% rubber powder.

By using this kind of classification, it is possible to express the results according the Figures 11 and 12. These figures show the strong restrictions for the design of a reactive compounding process:

- Low throughput
- Low barrel temperature
- High screw speed
- Medium filling rate

The intensity of kneading seems to be enough for different screw designs investigated.

Conclusions

Rubber powder can be dispersed using a twin screw extruder in-between a process length of approximately L/D = 14 including the feeding zone and the polypropylene melting.

A simultaneous feeding of both components into the hopper is possible even at high filling rates (70 wt.-% rubber powder).

The effective dispersion of the rubber powder enables a process design with a relatively long reaction time in the extruder. The whole process length of the used extruder is 44 L/D. It was shown, that the dispersion of rubber powder needs only around 14 L/D. A relatively long part of the process length (around 30 L/D) remains for reactive compounding, if the reactants are added downstream after the dispersion process.

Very high filling rates resulted in a lot of contacts between rubber powder particles, i.e. the system seems to be overfilled.

For design of a reactive compounding process a narrow processing window was observed.

[1] H. Michael, H. Scholz, G. Mennig, *Kautschuk Gummi Kunststoffe* 52 (**1999**) 7, p. 510-513J.
[2] H. Scholz, P. Pötschke, H. Michael, *Technomer'99* ,Chemnitz, 1999, proceedings C01/1-8
[3] S. Wiessner, H. Michael, G. Mennig, *Technomer'01*, Chemnitz, September **2001**, proceedings B12/1-10

Macromol. Symp. **2005**, *221*, 247-256

Dynamic Mechanical Behavior of PP/PET/MAPP Blends Filled with Glass Beads

M. A. Rodríguez-Pérez,[1] J. I. Velasco,[2] J. I. González-Peña,[1] J. L. Ruiz-Herrero,[1] D. Arencon,[2] J. A. de Saja[1]*

[1] Departamento de Física de la Materia Condensada, Facultad de Ciencias, Universidad de Valladolid, 47011 Valladolid, Spain
E-mail: marrod@fmc.uva.es
[2] Centre Català del Plàstic, Universitat Politècnica de Catalunya, Colom 114, 08222 Terrassa. Spain

Summary: PP/PET/MAPP blends have been filled with 50 wt% of glass beads. The orientation of the PP crystalline phase, the crystallization behavior and the dynamic mechanical response of these materials have been analyzed. The dynamic mechanical response is strongly affected by the presence of the glass beads, being possible to detect the effect of PET and MAPP on the storage modulus and loss factor values. Moreover, the alpha relaxation of the composites is visibly affected by thermal treatments.

Keywords: composites; differential scanning calorimetry; dynamic mechanical analysis; fillers; poly(propylene)

Introduction

Glass bead-filled polypropylene (PP) composites have a good processability, small and well-distributed internal stresses, good dimensional stability, and good service performance [1-3] . Due to these reasons several investigations have been performed for these kind of composites, focused on aspects related with the particle size, filler content, mechanical properties, interfacial adhesion, etc.[1-7]. However, taking into account the intricate micro-structure of these materials there are several aspects, as for example the dynamic mechanical behavior, in which more detailed studies seems to be necessary.

From a scientific point of view, it is very well known that particles with high aspect ratio are one of the main sources for the anisotropy of composites, examples of these materials are PP-talc, PP-aluminum hydroxide or PP-glass fibers [8]. One interesting feature of glass beads particles is their isotropy due to their spherical shape, and therefore they do not contribute to the composite anisotropy, so the induced polymer orientation due to the filler should also be reduced. It is also well known, that filler and polymer matrix orientation play an important role in the structure-

 DOI: 10.1002/masy.200550325

property relationships of a polymer filled system. In fact the effect of this parameter could cover the effect of other important structural aspect such as filler content, adhesion between filler and polymer, particle size distribution, etc. Therefore, it is expected that due to glass beads isotropy, more simple structure-property relationships could be obtained for the composites under study.

One of the most important aspects which modify a given polymer filled system is the interfacial adhesion between the filler and the polymer. In this investigation three different approaches have been considered to modify this feature. First, maleic-anhydride grafted polypropylene (MAPP) has been blended with the PP matrix, second glass particles have been treated with different coupling agents based on silanes, and finally poly(ethylene terephthalate) PET has also been added. It has been proposed the use of some polar thermoplastic polymers, such as polyamide [9] or polycarbonate [10], as efficient adhesion promoters in filled PP composites. Following this idea, we analyse here the use of poly(ethylene terephthalate-co-isophthalate) (co-PET) in glass bead filled PP.

Bearing the previous ideas in mind, in the present work, an attempt has been made to show how the dynamic-mechanical behavior of a series of PP-glass bead composites is affected by the addition of PET, maleic-anhydride grafted polypropylene (MAPP) and silane treated glass beads.

Materials

The polypropylene (Isplen PP050) used in this study was provided by Repsol-YPF. It is an isotactic homopolymer grade with a MFI (230 °C, 2.16 kg) of 5.0 dg/min. Glass beads with an average particle size of 20 μm were used as filler, being kindly provided by Sovitec Ibérica, S.A. A commercial grade of maleated polypropylene (EPOLENE G-3003), with an acid number of 8, was supplied by Eastman Chemical. Extrupet EW36 was the copolymer grade of PET manufactured by Catalana de Polímers S.A. (intrinsic viscosity of 0.8 dl/g) used in this investigation. Four different matrix compositions were selected as shown in Table 1a. Four different silane coupling agents were employed for surface tretament of the glass beads (Table 1b). The following procedure was used to homogeneously coat the glass bead with the silane: a solution containing 30 ml of silane, 250 ml of methanol, 60 ml of water and 5 ml of acetic acid was added to 1.5 kg of glass beads. This solution was stirred for 20 min to assure the hydrolysis of the alkoxy group in the silanes. The solution was transferred into a flask and glass beads were then added gradually while stirring. The mixture was then heated up to 40 °C to evaporate the solvent. Stirring was continued as long as the viscosity of the mixture was low enough. The non-reacted silane was washed out with methanol.

Composites containing nominally 50 wt% of glass beads and different matrix compositions were prepared using a Collin ZK-35 twin screw extruder. Standard dumb bell specimens (type "I" according to ASTM D638) were injection-molded using a 440/90 injection molding machine and a mould (tempered at 60°C) as described in Figure 4 (ASTM D-647 standard). In order to eliminate any residual stresses produced during injection molding the dumb bell test specimens were annealed at 110 °C for 24 h.

Table 1a.Composition of the matrix expressed in percentages by weight.

Reference	A	B	C	D
Matrix composition	PP (100)	PP/MAPP (97/3)	PP/PET (95/5)	PP/MAPP/PET (92/3/5)

Table 1b. Silane treatments applied on the glass bead surfaces. Supplied by a) Dow Corning, b) Witco

Surface treatment reference	**Silane**	**Commercial name**
1	–	–
2	3-metacriloxipropyl trimethoxy silane	Z-6030[a]
3	3-mercaptopropyl trimethoxy silane	A-189[b]
4	N-(2-aminoethyl)-3-aminopropyl trimethoxy silane	Z-6020[a]
5	N-(2-(vinylbencylamino)-ethyl)-3-aminopropyl trimethoxy silane	Z-6032[a]

Experimental

Wide angle X-ray diffraction experiments (WAXD) were performed to analyze the PP crystalline orientation. The experiments were performed using a Phillips PW 1050/71 diffractometer. Radial scans of intensity (I) versus scattering angle (2θ) were recorded in the range 5-50 ° by using filtered Cu K_{α} radiation. The testing was carried out using prismatic bars, extracted from the dumb bell specimens (Figure 1), of nominal dimensions 25 x 5 x 3 mm^3.

Differential scanning calorimetry (DSC) measurements were performed using a Perkin-Elmer Pyris 7 calorimeter. Sample mass was between 8 and 9 mg, the samples were cut from dumb bell specimens as showed in Figure 1. After the erasing of the thermal history (200°C for 4 minutes) cooling measurements were conducted from 200 to 25 °C with several cooling rates: 5, 10, 15, 20, 25, 30, 40 and 60 °C/min. All runs were carried out in a stream of dried nitrogen. After each cooling experiment a heating run between 25 and 200 °C was performed at 10 °C/min.
The method developed by Dobreva and Gutzow [11] was applied to analyze the non-isothermal crystallization experiments. From this analysis, a nucleating activity parameter (ϕ) can be obtained for the filler. The value of ϕ decreases from 1 to 0 as the polymer crystallizes in the presence of more efficient nucleating particles. This approach has been successfully applied to evaluate the nucleating rate of PP filled with different mineral particles [12-14].

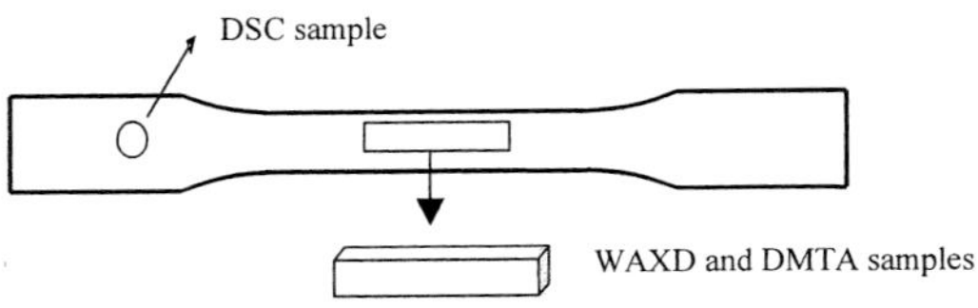

Figure 1. Injection-molded dumb bell specimen showing the sample extraction zone for DSC analysis and the machined prismatic specimen used in WAXD and DMTA analysis.

Dynamic mechanical thermal analysis (DMTA) was carried out using a Perkin-Elmer DMTA 7 and a Mettler DMA861-e equipments. The testing configuration was three-point bending with a span of 20 mm. The samples employed were the same used for WAXD measurements. A static stress of 6 MPa and a dynamic stress of ± 5 MPa were applied with a frequency of 1 Hz. For each material two kinds of tests were performed. First, experiments at 20 °C were performed to obtain quantitative results of the storage modulus (E') and the loss factor (tan δ). The values were collected 5 min after applying the stresses, i.e., once the initial fluctuations of the values had disappeared. Second, the study of the composites relaxations was carried out by tests performed in the temperature range from –40 to 130 °C at a heating rate of 5 °C/min.

Results

The orientation of the monoclinic α-phase PP crystals in the injection-molded samples has been studied by WAXD. The relationship between the orientation of PP a- and b-axes can be determined by calculating the ratio of the intensity of the PP (040) plane (peak at 2θ=16.7 ° in the diffraction pattern) to that of the (110) plane (peak at 2θ=13.9°). The results obtained are summarized in Table 2. Several values of this ratio can be found in the literature for filled polypropylene [12,14]. Moreover, several authors have reported values for an isotropic mixture of PP crystallites, for example 0.54 [15] and values comprised between 0.67 and 0.77 [16].

Table 2. WAXD results, activity parameter (Φ), storage modulus (E') and loss factor (tan δ), glass transition temperature (T_g) and α relaxation temperature (T_α).

Sample	$\frac{I(040)}{I(110)}$	Φ	E'(GPa)	tan δ	T_g (°C)	T_α (°C)
PP	1.44	1.00	1.6	0.054	-0.8	52.4
A1	0.98	0.72	3.0	0.041	-1.1	56.1
A2	1.06	0.77	3.5	0.040	-0.5	61.6
A3	1.00	0.62	3.2	0.039	-1.4	61.1
A4	0.97	0.71	2.9	0.035	0.5	61.1
A5	0.90	0.66	3.2	0.030	0.6	59.4
PP/MAPP	—	1.00	—	—	—	—
B1	1.17	0.53	2.8	0.040	0.2	57.6
B2	0.97	0.46	3.1	0.039	0.2	61.6
B3	1.10	0.55	3.2	0.035	0.2	62.5
B4	1.21	0.52	3.3	0.035	0.9	61.7
B5	1.07	0.55	3.0	0.032	0.9	61.7
PP/PET	1.17	0.52	1.7	0.046	0.3	52.1
C1	1.05	0.49	3.1	0.034	0.1	57.1
C2	1.01	0.56	3.1	0.032	0.0	59.6
C3	0.99	0.53	3.0	0.040	0.8	59.8
C4	0.95	0.46	3.0	0.031	1.0	65.0
C5	0.93	0.49	3.3	0.029	1.2	61.9
PP/MAPP/PET	1.35	0.51	1.9	0.042	0.7	54.6
D1	1.08	0.52	3.4	0.032	0.6	64.5
D2	1.20	0.46	3.2	0.031	0.6	64.3
D3	1.07	0.52	3.2	0.030	1.0	59.4
D4	1.06	0.36	3.4	0.028	1.3	62.6
D5	1.02	0.41	3.4	0.031	1.3	59.2

It is noticed that both PET and glass beads lead to intensity ratios close to 1, resulting in an homogenization of the polypropylene orientation due to these two components. The PET effect contrasts with that observed by Saujanya et al. [17] in PET fiber-reinforced PP composites in which the crystallite growth along b-axis is strongly affected by the presence of PET fibers. It should be mentioned that substrates with higher aspect ratio usually involve transcrystallinity structures in polypropylene [18, 19], allowing a preferential location of polypropylene chains. In previous investigations [12, 14] the anisotropy ratios for other PP systems were measured, the value was 6 for PP-talc composites, 21 for PP-silane treated talc materials, and 19 for PP-magnesium hydroxide materials. Therefore, as it was assumed in the introduction of this paper, glass beads particles, in comparison with other fillers, reduce the PP crystalline phase orientation of the composites.

On the other hand, it was appreciated that MAPP slightly promoted PP orientation, probably, due to the compatibilizing action of the copolymer. Finally, silane treatments does not have an obvious effect on the PP orientation.

The nucleating activity (ϕ) of glass beads is not to high (0.70) in comparison with that of other fillers (see DMTA results). The activity parameter is smaller (0.52) when PET was added to PP, indicating a higher nucleating rate. MAPP presence enhanced the nucleation activity of glass beads, leading to a significant reduction in ϕ values for all the composites with PP/MAPP matrix. On the other hand, glass beads addition into PP/PET and PP/MAPP/PET blends did not greatly change the values of ϕ when compared with unfilled blends.

With respect to the surface treatments applied on glass beads, although controversial opinions are found in the literature, in this work it was detected that the use of silanes with any aminofunctional group (silanes Z-6020 and Z-6032) enhanced slightly the nucleation activity of the glass beads.

Storage modulus values obtained from isothermal test (Table 2) showed, in unfilled samples, that the addition of PET increases slightly the stiffness. This increase is higher when MAPP was added, probably due the compatibilizing effect of MAPP in the PP/PET blend. The presence of glass beads in the PP matrix increases noticeably the stiffness of the compound, although there is no clear trend concerning the effect of the silane surface treatments applied.

It was observed a slight decrease (Table 2) in the value of the loss factor when PET was added to the polypropylene matrix. At room temperature, PET is under its glass transition temperature, therefore, a reduction of the viscoelastic character of the material is expected. A further reduction of the loss factor value was observed in the PP/MAPP/PET blend. This fact has been related to an enhanced interfacial adhesion between phases [10], which would hinder the molecular motion of polypropylene chains at the interface.

It is observed a remarkable reduction of the tan δ value when PP is filled with 50 wt% of untreated glass beads. This behavior is due, on the one hand, to the lower weight fraction of polymer in the composite, and on the other hand to the reduced mobility of the polymer phase close to the glass particles. Surface treatment of glass beads leads to a further reduction, in an extent that depends on the silane employed. Once again, this result should be understood in terms of the reduced mobility of the polymer molecular segments at the interface.

The loss modulus (E'') for all the samples has a maximum around 0 °C (Figure 2), which is associated with the glass transition of polypropylene (T_g). The obtained values for the glass transition temperature are collected in Table 2. Small differences between samples were established, as the values ranged between –1 and 1 °C. Moreover, there is no clear trend concerning the effect of glass beads. Nevertheless, as a general trend, it was observed that all the composites containing MAPP showed slightly higher T_g values than their homologous without copolymer. In addition, the highest values for each matrix composition were displayed by the samples with glass beads treated with silanes Z-6020 and Z-6032. This slight increase should be related with stronger interactions between matrix and filler promoted by silanes.

The shifting of the glass transition when PP is filled is an interesting subject in which a different behavior has been observed depending of the used filler. For instance, a reduction of the T_g value [12] (measured by DMTA) was observed for both PP-talc and PP-silane treated talc composites. The results for PP-Mg $(OH)_2$ or PP-Al $(OH)_3$ composites are similar to that for glass beads [13,14], the T_g remains constant. In our opinion, one of the main reasons for these differences is the different nucleating activity of each filler in the polymer crystallization. The nucleating activity of glass beads into PP (ϕ=0.70) is less effective than that displayed by crystalline mineral fillers like talc (ϕ=0.32 [12]), magnesium hydroxide (ϕ=0.52 [13]) or aluminum hydroxide (ϕ=0.54 [14]). Therefore, a very high nucleating activity seems to modify the PP morphology in an extend which shifts the glass transition temperature.

Both unfilled and filled samples displayed an additional maximum (T_α) located around 50-70 °C (Figure 2). Controversial opinions are found in the literature about this relaxation. McCrum [20] ascribes this relaxation to both mechanisms of lamellar slipping and rotation between crystalline entities. Jančář [21] supposes that this maximum is the consequence of a release of strongly hindered segmental mobility of molecules from the interface on the filler surface. This author also supposes that the relaxation is probably the glass transition peak of PP immobilized close to the filler surface, reporting the development of this relaxation with filler contents above 26 % vol., while it was no detected for unfilled PP. On the other hand, Stricker et al. [2] ascribe the existence of this transition to the quenching melts of glass bead filled PP, noticing that slow cooling and crystallization avoid this effect and discard the existence of the maximum due to presence of immobilized matrix.

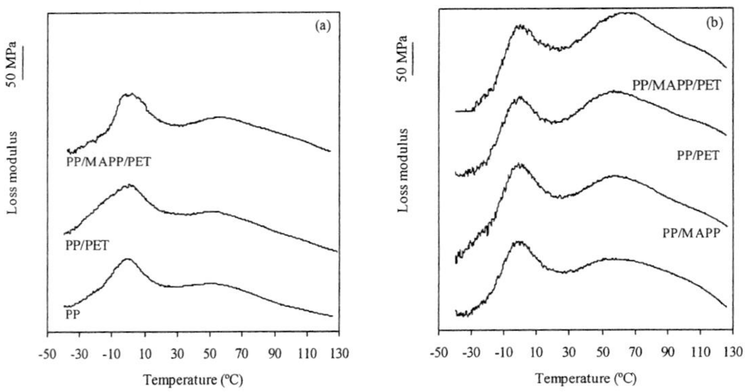

Figure 2. Loss modulus vs. temperature of a) unfilled samples and b) composites with untreated glass beads. Heating rate = 5 °C/min. Frequency = 1 Hz.

The values of the maximum of the α relaxation (T_α) are collected in Table 2. Contrary to the results observed by Jančář, we have observed this relaxation in the unfilled samples. Both its intensity and temperature have been very similar for the polypropylene and the PP/PET blend (Figure 2). On other hand, in the blend PP/MAPP/PET, the relaxation was more intense and shifted to higher temperatures, which might be a consequence of the compatibilizing effect exerted by MAPP. In the filled composites, the α relaxation showed the highest intensities, detecting that PET presence causes a build-up of the intensity. It is perceptible that in general terms silane surface treatments increased the T_α values.

From the obtained results it seems that this relaxation depends on the interactions between the PP and the other phases, increasing its intensity and shifting to higher temperatures when both glass beads and PET are added. Moreover, a similar trend was observed when MAPP and a glass bead surface-treated with silane were included in the formulation. Nevertheless, the existence of this relaxation in the unfilled PP involves a relationship with the PP crystalline phase, and may suggest that the special molding conditions (mould temperature 60 °C, cooling time 30 s, annealing at 110 °C for 24 h) applied to the produced materials could induce a particular crystalline structure which is able to develop into more stable states, under the DMTA testing conditions.

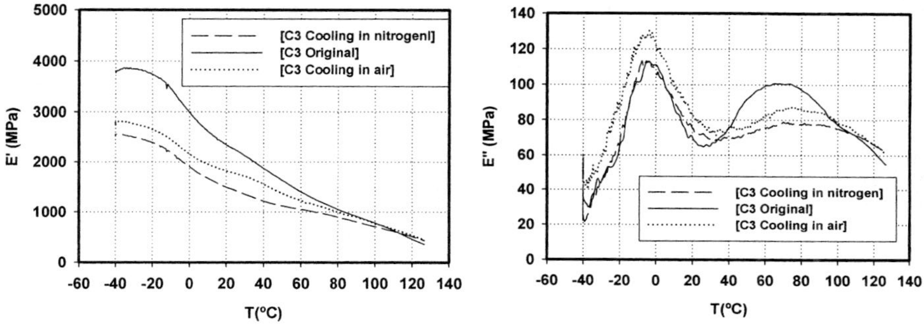

Figure 3. Dynamic Mechanical behaviour of sample C3 under different thermal treatments.

To check the previous hypothesis, the thermal history of several samples was modified by performing two different thermal treatments. First, annealing at 145° C for 1 hour and cooling in liquid nitrogen and second annealing at 145 °C for 1 hour and cooling in air. After these treatments, DMTA experiments in a Mettler 861-e equipment were carried out. A typical example of the results (C3 sample) is shown in Figure 3. As it can be observed, there is an obvious difference between the original material and the treated ones. Due to the treatments the storage modulus is clearly reduced, the alpha transition is shifted to higher temperatures and its intensity is also reduced. These two effects are more marked for the sample cooled in liquid nitrogen. Moreover, the glass transition temperature does not change by the treatments. Therefore, the treatments modify the crystalline structure of the polymer matrix and this modification is detected by DMA in a softer material, showing a less intense alpha relaxation.

Conclusion

Storage modulus of polypropylene increased slightly with the addition of PET, and considerably when glass beads were added, whereas the loss factor followed the opposite trend. Some of the DMTA trends seemed to be related to the compatibilizing effect of MAPP. Very small differences in the glass transition temperature of both unfilled and filled samples were found, probably due to the low activity of the filler. Under the applied processing conditions, it has been observed the α relaxation of PP at temperatures between 50-70 °C. This relaxation has been detected for all the materials under study, being its intensity and peak temperature determined by the presence of PET, MAPP and glass beads. It has also been observed that this relaxation is very sensitive to thermal treatments.

Acknowledgements

Authors thank the MCYT (Government of Spain) for the financial support of MAT-2000-1112 and MAT-2000-0084 P4 projects. D. Arencón thanks the CIRIT (Government of Catalonia, Spain) for the concession of a predoctoral grant.

[1] F. Stricker, M. Bruch, R.Mülhaupt, *Polymer.* **1997**, 38, 5347-5353.
[2] F. Stricker, R.D. Maier, Bruch M, Thomann R, Mülhaupt R, *Polymer.* **1999**, 40, 2077-2084.
[3] J.Z. Liang, R.K.Y Li, *Polym Int,* **2000**, 49, 170-174.
[4] J.Z. Liang, R.K.Y. Li Tjong SC, *Polym Testing.* **2000**, 19, 213-220.
[5] D.L Faulkner, L.R. Schmidt. *Polym Engng Sci.* **1977**,17, 657-665.
[6] L.R.Schmidt. *Polym Engng Sci.* **1977**, 17, 666-670.
[7] J.Z., Liang, R.K.Y. Li . *Polym Compos* **1998**, 19, 698-703.
[8] D.Bikiaris,P.Matzinos,A.Larena,V.Flaris, C.J.Panayiotou. *J Appl Polym Sci.* **2001**, 81, 701-709.
[9] M. Ulrich, C. Caze, P. Laroche, *J Appl Polym Sci.* **1998**, 67, 201-208.
[10] H. Noh, B.S. Yoon, M.H. Suh, S.H. Lee, *Polymer.* **2001**, 42, 2695-2700.
[11] A. Dobreva-Veleva, I. Gutzow, *J Non Cryst Solids*, **1993**, 162, 13-25.
[12] M. Alonso, J.I. Velasco, J.A. de Saja, *Eur Polym J.* **1997**, 33, 255-262.
[13] J.I. Velasco, C. Morhain, A.B. Martínez, M.A. Rodríguez-Pérez, J.A. de Saja, *Macromol Mater Eng.* **2001**, 286, 719-730.
[14] J.I. Velasco. C. Morhain, A.B. Martínez, M.A. Rodríguez-Pérez, J.A. de Saja, *Polymer.* **2002**, 43, 6813-6819.
[15] E.J. Addink, J. Beintema, *Polymer.* ***1961***, 2, 185-190.
[16] F. Rybnikar, *J Appl Polym Sci.* **1989**, 38, 1479-1490.
[17] C. Saujanya, S. Radhakrishnan, *Polymer*, **2001**, 42, 4537-4548.
[18] J.K. Tan, T. Kitano, T. Hatakeyama, *J Mater Sci.*, **1990**, 25, 3380-3384.
[19] J.L. Thomason, A.A. Van Rooyen, *J Mater Sci.*, **1992**, 27, 889-896.
[20] McCrum, E. Read BE, Williams G. *"Anelastic and dielectric effects in polymeric solids"*. Wiley, London, **1967**.
[21] J. Jančář, *J Mater Sci.* **1991**, 26, 4123-4129.

Modified Cellulose Fibers as Reinforcing Fillers for Macromolecular Matrices

*Alessandro Gandini, Mohamed Naceur Belgacem**

Ecole Française de Papeterie et des Industries Graphiques (EFPG), BP 65, 38402 Saint Martin d'Hères, France
E-mail: naceur.nelgacem@efpg.inpg.fr

Summary: Various cellulose substrates were submitted to different surface OH chemical modifications in non-swelling media and then fully characterised by FTIR, XPS, contact angle measurements and elemental analysis. The alternative procedures adopted were:

(i) simple grafting with monomeric or oligomeric agents which bore one or several reactive functions, either as a single terminal moiety or as numerous moieties placed along the chain;
(ii) grafting with polymerisable molecules, which involved the use of reagents of small molecular size, but bearing two functions, one able of reacting with the surface cellulose OH groups and the other used as a source of subsequent covalent linkage with the matrix;
(iii) the use of planar stiff molecular configurations bearing two identical reactive functions, calling upon the working hypothesis that only one function will react with a cellulose OH group, whereas the other will be left to copolymerise with the matrix;
(iv) coupling with siloxanes using the same strategy as in (ii), except for the fact that this specific approach involves molecules already commonly used in glass fibres treatment;
(v) grafting with organometallics such as triethyl aluminium following the same working hypothesis as in (iii) but with a second reaction involving monomeric or polymeric alcohols or amines.

Keywords: cellulose; chemical grafting; surface characterisation; surface treatment

Introduction

The chemical modification of the surface of lignocellulosic fibres has attracted considerable attention in recent years because of the growing interest in the use of these natural filaments as reinforcing agents in macromolecular composite materials[1]. The main driving force of this

 DOI: 10.1002/masy.200550326

research activity resides in the possibility of recycling such composites at the end of their life cycle, in contrast with glass-fibre-based counterparts. Additional advantages related to the use of lignocellulosic fibres include their renewable character, and the fact that they are available ubiquitously at modest costs, albeit from different sources like annual plants and wood species.

The quality of the fibre-matrix interface in composite materials plays a fundamental role on their mechanical properties and often requires the surface modification of the fibres to achieve maximum compatibility and therefore good adhesion. The functions exploited for these coupling reactions applied to cellulose include isocyanates, anhydrides, oxiranes and siloxanes, i.e. moieties capable of reacting with the surface hydroxyl groups of the fibres. Additionally, the modification should provide an efficient hydrophobic barrier for the cellulose fibres in order to avoid moisture uptake and the consequent loss of mechanical properties, in the event of spurious contacts with ambient air (e.g. from the open end of a sectioned piece of the composite material). Our laboratory has been actively involved in all these strategies[2-10], as well as in a more "physical" alternative, which calls upon the construction of a sleeve of polymer around the fibres by admicellar polymerisation[11]. The purpose of this paper is to provide a broad outlook of these different chemical approaches, as applied in our laboratory, with particular emphasis on recent advances.

Experimental

Several types of cellulose substrates were used for these modifications, including a semi-crystalline powder (Avicell), a home-made additive-free tracing paper (NMMO-regenerated continuous fibres with a regular diameter of about 30 μm (NMMO = N-Methyl-Morpholine) and Whatman filter papers. This allowed the various techniques of characterisation described below to be applied to the appropriate corresponding morphology. All reactions were carried out heterogeneously with the fibres suspended in the organic solution of an excess of the grafting reagent. The temperature and duration of the modifications depended on the specific context and varied between room temperature and 80°C and between a few hours and a few days, respectively. The characterisation of the modified fibres always followed the removal of the unreacted agent, by a Soxhlet extraction with the appropriate solvent. FTIR in different modes (Transmission, ATR, DRIFT) and XPS spectroscopy, elemental analysis, dynamic and static contact angle measurements, inverse gas chromatography and scanning electron microscopy (SEM) were the essential tools utilised to assess the occurrence and the extent of the modification.

Results and Discussion

The different strategies and the corresponding results are discussed separately below, before drawing comparisons about their efficiency and potential applications. References are given to studies which have already been published for the reader who would wish to have more detailed information.

Simple grafting [2-4]

The fibres were treated with a variety of reagents bearing one or several functional groups (NCO, anhydride,...) in a cellulose non-swelling medium, e.g. methylene chloride or tetrahydrofuran (THF). Some of these agents were oligomeric structures which bore one reactive function, either as a terminal moiety (so that the coupling would give rise to "hairs" on the fibres), or as a central moiety along the oligomer (so that "umbrellas" would be clipped on the fibre surface). The modification took place successfully, albeit in modest amounts, as expected by the fact that only the surface OH groups could react in these conditions. A typical example of this approach is the reaction of a commercial alkyl succinic anhydride (ASA, structure 1), currently used as sizing agent in papermaking, viz. a mixture of C10 to C14 homologues bearing one cyclic carboxylic anhydride function pending from the middle of each molecule (umbrellas). After extraction of the excess of ASA, the fibres displayed a strong water repellence and the polar contribution to their surface energy decreased considerably, compared with that of the starting fibres.
Another type of reagent used here consisted of copolymers containing both conventional and reactive monomer units, so that the multiple coupling reactions between the cellulose OH groups and at least some of the functional groups of each macromolecule would append bridge-like structures at the surface of the fibres. As an example, a random poly(α-methyl-styrene-*co*-3-isopropenyl-α,α'-dimethylbenzyl isocyanate) (structure 2) was synthesised by cationic initiation. The ensuing NCO-modified styrene-like copolymer was carefully characterised in terms of structure and molecular weights by Vapour Pressure Osmometry and was found to have a DP of

~200 and contained some 20 NCO function per copolymer molecule[12, 13]. Its reaction with the cellulose fibres was monitored, after extraction of the unreacted copolymer, by elemental analysis (nitrogen content), FTIR (typical peaks for the urethane C=O and the aromatic moieties), XPS (presence of nitrogen and modification of the oxygen peaks of cellulose).

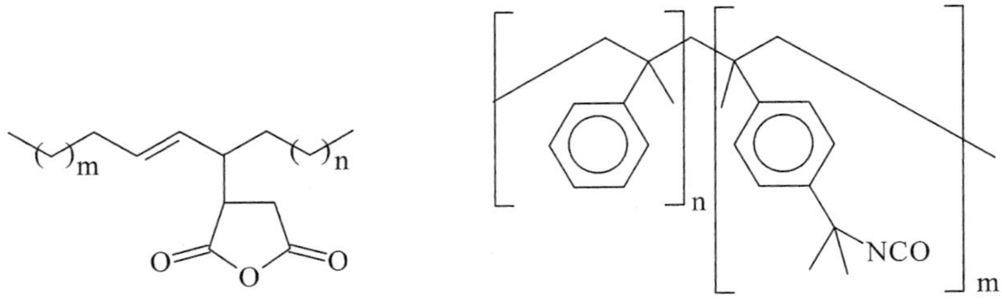

Structure 1

Structure 2

However, the most significant changes appeared in the fibre surface properties, as shown by contact angle measurements, which led to the calculation of the dispersive (34 mJ m^{-2}, as with polystyrene) and polar (only 1 mJ m^{-2}, compared with ~25 mJ m^{-2} for the starting cellulose) contributions to the surface energy, and by the observation of the polymer film covalently attached to the fibres by electron microscopy. Figure 1 shows an example of the contact angle of drop of water before and after grafting with polymer 2. The contact angle on the untreated paper rapidly decreased by spreading and absorption, whereas that related to the modified surface remained unchanged for hours.

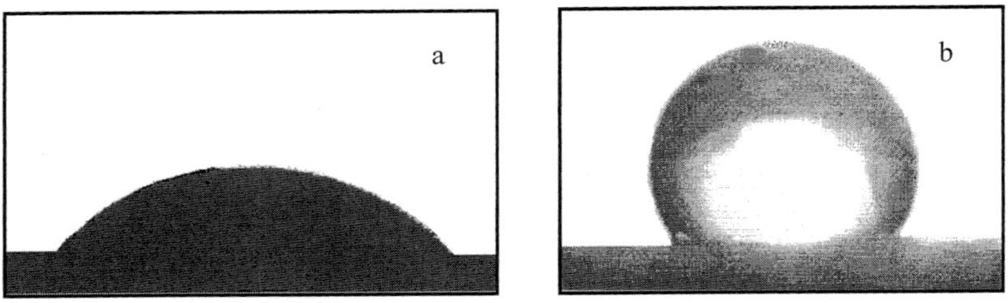

Figure 1. Instantaneous contact angle of a drop of water on tracing paper samples (a) before, and (b) after grafting with polymer 2.

Figure 2 gives the SEM microphotographs of NMMO cellulose fibres before and after modification with that coupling agent. Table 1 summarises the surface energy of cellulose before and after grafting with different agents, including those mentioned in this section.

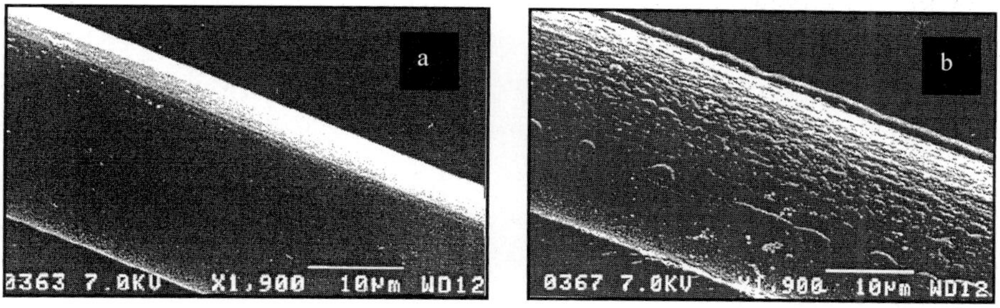

Figure 2. SEM micrographs of a NMMO cellulose fibre (a) before, and (b) after grafting with polymer 2.

Grafting with polymerisable molecules [2,3,5]

The next step in this saga called upon a different approach, which involved the use of reagents of small molecular size, but bearing both a function capable of reacting with the surface cellulose OH groups and one to be used as a source of covalent linkage with the matrix. Structures 3 to 6 illustrate this point:

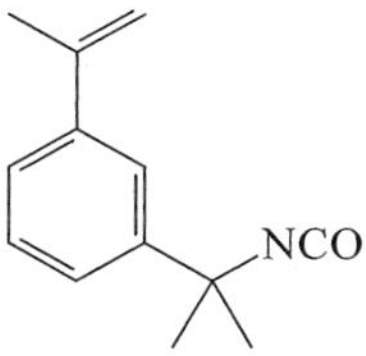

Structure 3

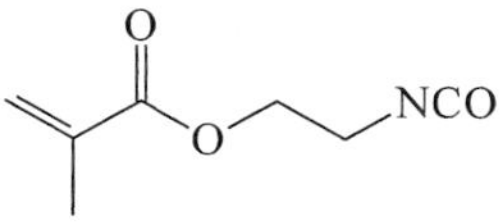

Structure 4

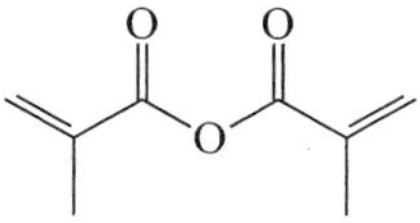

Structure 5

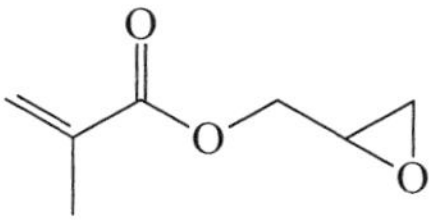

Structure 6

The procedure adopted with these reagents consisted in (i) carrying out the grafting with the fibres suspended in the monomer which was chosen as the source of the macromolecular matrix and thereafter (ii) induce the polymerisation of the medium. The polymerisable styrenic or acrylic moieties, attached to the surface of the fibres following the first grafting reaction, were supposed to participate in the polymerisation of the bulk monomer as co-monomer units and thus provide a continuity of covalent bonds between the substrate and the matrix. This was confirmed by extracting the entire unbound polymer and characterising the modified fibres. Their weight gain was as high as 10-15% (to compare with 1-2% when simple coupling with large molecules described in previous section) and the presence of covalently anchored polymer was readily detected by FTIR spectroscopy.

Thus FTIR spectroscopy was used to ascertain the occurrence of the grafting at each step and Figure 3 give a comparison of the FTIR spectra of microcrystalline cellulose (avicel) before, after treatment with **4,** and after copolymerisation with methyl methacrylate (MMA) and then extracted to remove all unbound polymer. This figure gave clear-cut evidence about the occurrence of the first step of grafting, by the presence of two main bands, namely at 1715 and 1642 cm^{-1} associated with the formation of ester moiety and the C=C of **4**, respectively (Figure 3b). Further evidence on the occurrence of the second step of grafting (copolymerisation between the growing MMA macroradicals and the unsaturation of **4**) can be shown examining Figure 3c. In this FTIR spectrum one can observe the increase of the band at 1726 cm^{-1}, attributed to carbonyl groups, which moreover shifted to higher frequencies. This is a clear indication of the presence of polyMMA chains bounds to the cellulose surface. The appearance of a shoulder at 2955 cm^{-1} associated to methyl groups of polyMMA confirms the above given hypothesis.

As expected, the surface properties of the fibres modified by this approach, determined by contact angle measurements applying the classical Owens-Wendt equation, resembled very closely those of the polymer which had been bound around them (Table 1). Apart from the actual success of the desired modification, the additional interest of this operating procedure is the fact that no solvent is needed, since the one-pot operation provides both the grafting of the fibres and the composite manufacture. The excess of reagent used in the first step is also utilised, since it polymerises with the matrix monomer during the second and final stage. It is important to emphasise that the presence of continuous covalent links between the fibres and the matrix represents the best way to optimise the quality of the adhesion at the interface between the two components of the composite. Examples of the composite materials prepared following this strategy include matrices made of polystyrene and different acrylic polymers.

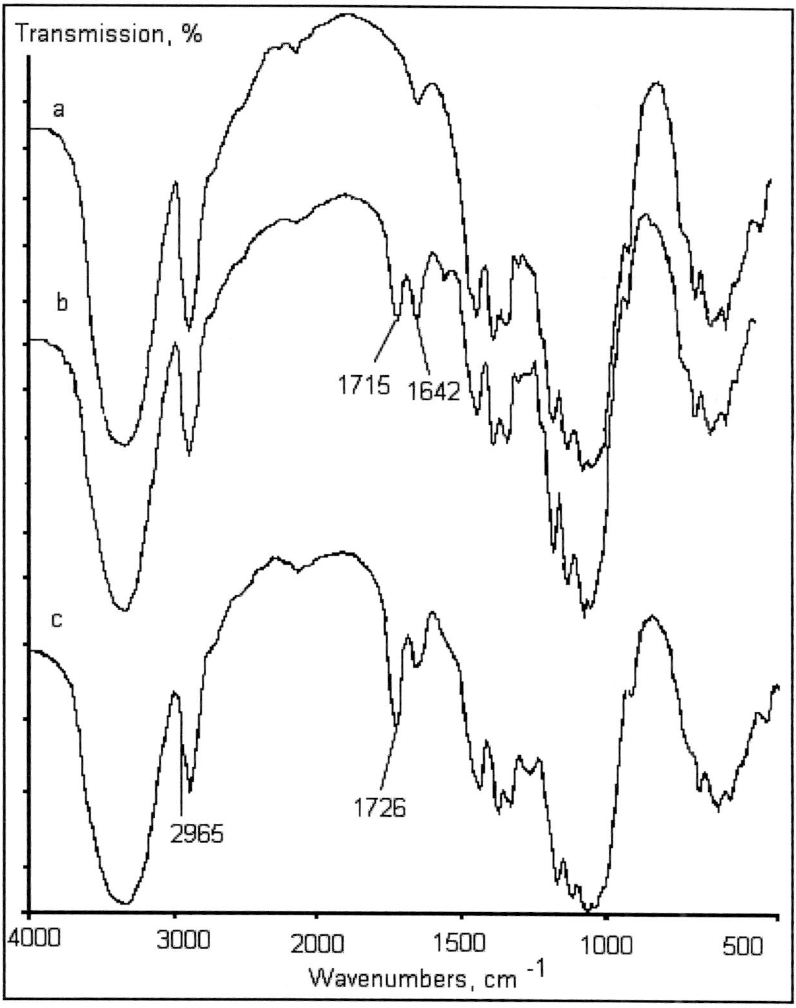

Figure 3. Transmission FTIR spectra of avicel before (a), after treatment with 4 (b), and after copolymerisation with methyl methacrylate (c).

Table 1. Surface energies of un-sized additive-free tracing paper before and after different treatments.

Grafting		Surface energy, mJ/m^2			Reference
First	Second	γ^D	γ^P	γ^T	
None	None	29.4	20.0	49.4	[4]
structure 1	None	27.3	8.6	35.9	[4]
structure 2	None	34.4	1.0	35.4	[4]
structure 3	None	42.7	0.3	43.0	[5]
structure 4	None	36.7	0.8	38.5	[5]
structure 3	Styrene	30.9	0.8	31.7	[5]
structure 3	MMA	44.6	0.1	44.7	[5]
structure 4	Styrene	30.9	0.8	31.7	[5]
structure 4	MMA	39.1	0.1	39.2	[5]
structure 9	None	46.9	0.1	47.0	[9]
structure 9	MMA	24.9	1.3	26.2	[9]
structure 9	Styrene	36.6	0.1	36.7	[9]
structure 10	None	33.3	9.2	43.5	[9]
structure 10	Epoxy	37.9	0.8	38.7	[9]
structure 12	None	30.5	6.3	36.8	[9]
structure 13	None	32.0	4.0	36.0	[9]
structure 13	Styrene	40.3	2.3	42.6	[9]
structure 13	None	43.5	0.8	44.3	[9]
BEt_3	1-heptanol	29.0	8.1	37.1	[10]
BEt_3	4-ABN	31.0	15.5	46.5	[10]
BEt_3	PFPEO	21.0	7.5	28.5	[10]
$AlEt_3$	PFPEO	5.8	1.5	7.3	[10]

MMA: methyl methacrylate, ABN: aminobenzonitrile, PFPEO: α-hydroxy-perfluorated polyethylene oxide.

The use of reagents with two identical reactive functions [6].

Whereas in the preceding section the structure of the grafting molecules involved two specific moieties, one for each step in the processing of the composite, here the working hypothesis was inspired by the use of molecules bearing two identical reactive moieties. These structures were chosen on the basis of specific considerations which called upon planar rigid configurations with the two functions in "opposite" directions, as shown in structures 7 and 8.
It was thought that, given the heterogeneous character of the reactions between the cellulose surface OH groups (solid) and the complementary moieties borne by these reagents (in the liquid phase), only one of the two moieties would be able to couple with the fibre, leaving the other available for further exploitation. This hypothesis was confirmed experimentally, when cellulose was treated with either of the above reagents in a dry non-swelling medium, sine the recovered fibres showed, after extraction of the excess coupling agent, the clear presence of the relevant infrared bands arising from the NCO or anhydride carbonyl groups, as shown in Figure 4 in the case of the former. In fact, the intense band at 2270 cm^{-1} denotes the presence of NCO groups attached to the substrate (see Figure 4). In other words, the use of these specific molecules provided a simple and convenient means to prepare cellulose fibres bearing highly reactive groups bound at their surface.

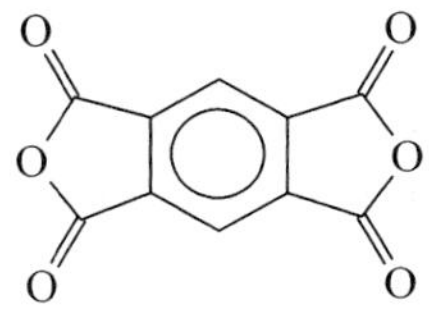

Structure 7

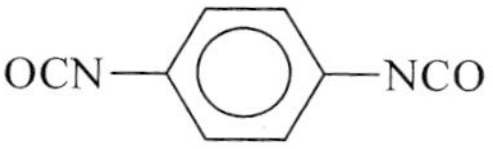

Structure 8

The exploitation of this type of modification consisted in carrying out, once more, the processing of the composite material through a one-pot double operation, viz. (i) the modification of the surface of the fibres followed by (ii) polycondensation reactions in which the unreacted function could participate and thus give rise to continuous covalent links between fibres and matrix. Examples of these procedures included the successful preparation of composites with polyester (with anhydride-bearing fibres) and polyurethane (with NCO-bearing fibres) matrices. Following the extraction of the polymer which had not "copolymerised" with the cellulose reactive groups, the modified fibres showed weight gains of up to 200% and were thoroughly embedded in the corresponding polymer, which was chemically bound to them.

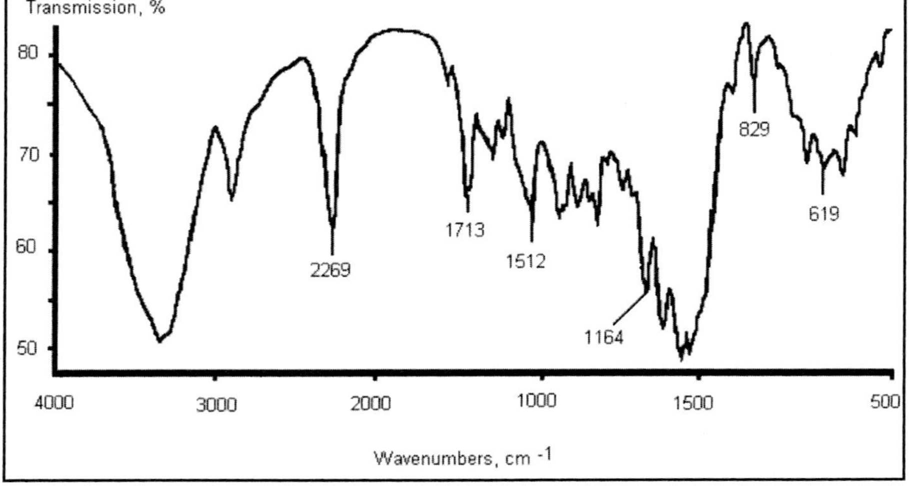

Figure 4. ATR-FTIR spectrum of cellulose grafted by 1,4-phenylene diisocyanate (structure 8) after extraction with dry methylene chloride.

Coupling with siloxanes [7-9]

Whereas the reaction of siloxanes with glass fibres is very well documented, the same cannot be said of the corresponding coupling with cellulose fibres, because claims to the actual occurrence of these grafting reactions are often unsubstantiated by convincing experimental evidence. To the best of our knowledge, the only study which approached this problem in a thorough fashion was published by Herrera-Franco *et al.* [14]. The issue at stake here is double, namely the actual reactivity of the cellulose hydroxy groups with either the ≡Si-OR or the ≡Si-OH moieties, the first being associated with unhydrolysed siloxanes and the second with their hydrolysed oligomers. Herrera-Franco *et al.* were the first to give a clear answer to the second aspect: the reaction ≡Si-OH + ≡C-OH (solid cellulose) only takes place at reasonable rates if the temperature is higher than about 70°C.

Our recent work has extended that study in various directions and can be summarised as follows: (i) if triethoxysilanes are prehydrolysed to give essentially ≡Si-OH-bearing oligomers, strong hydrogen-bond interactions take place with the surface hydroxy groups of cellulose at room temperature, but the silanol oligomers can be removed by washing with alcohol; (ii) if the same system is heated above 100°C, a condensation reaction takes place to give ≡Si-O-C≡ bonds; (iii) if monomeric siloxanes are placed in contact with cellulose fibres in a dry solvent (e.g. toluene), no reaction takes place, even at high temperature, unless moisture is introduced in the medium. It can therefore be concluded that the chemical modification of cellulose fibres by siloxanes must be mediated by their preliminary (partial) hydrolysis and even then it will only occur to a measurable extent if the temperature is raised.

A selection of trialkoxy-silanes (structures 9-14) bearing different functions on the fourth substituent (including C=C unsaturations) were examined quantitatively in terms of adsorption isotherms and coupling reactions with cellulose fibres. The fibres modified with siloxanes possessing a copolymerisable function were shown to give covalent connections with a macromolecular matrix when the latter was prepared *in situ* from the polymerisation of its monomer. Table 1 gives the surface energies of a selection of silane-treated celluloses as well as silane-treated and copolymerised fibres. Figure 5 shows the evolution of the contact angle values with short times of a drop of water placed on the cellulose surface modified by different silanes.

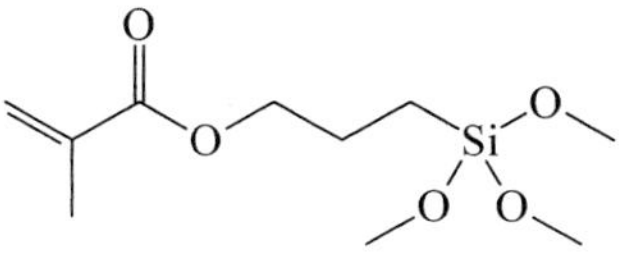

Structure 9

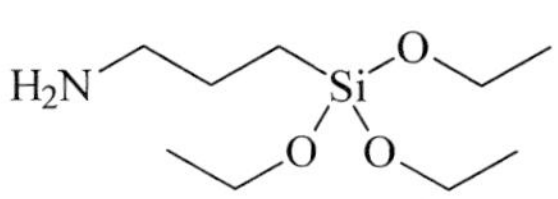

Structure 10

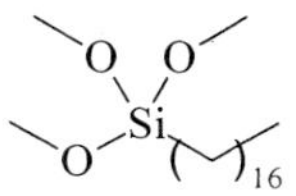

Structure 11

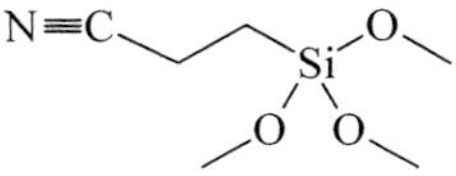

Structure 12

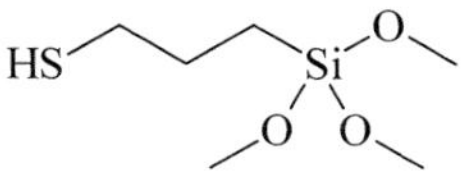

Structure 13

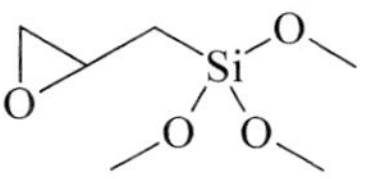

Structure 14

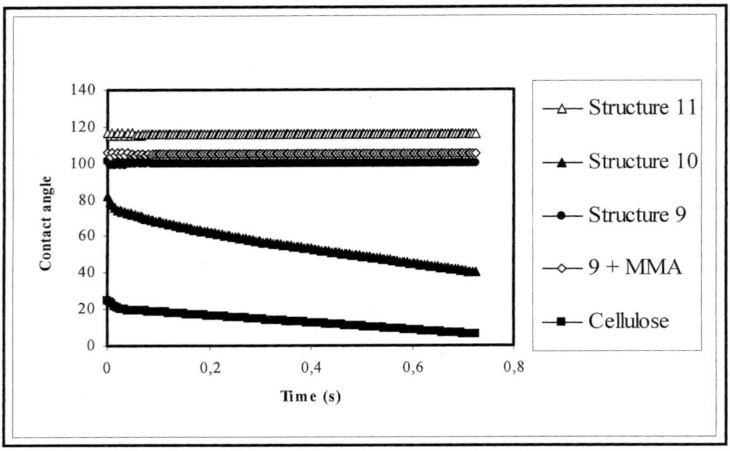

Figure 5. Contact angle kinetics of a drop of water placed on the surface of cellulose modified by different silanes.

Grafting with organometallics [3,10]

Among our most recent investigations in the context of the present topic, we tackled the reaction of cellulose fibres with organometallic derivatives of aluminium and boron. To the best of our knowledge, no study of this nature had been published previously. The fibres were treated with triethyl aluminium (Et_3Al) or triethyl boron (Et_3B) in a dry heptane medium at room temperature under a nitrogen atmosphere. After filtering off the excess reagent and washing in situ with a dry solvent, a second reaction was carried out, involving the addition of one of several alcohols or amines. If one or two of the metal-to-carbon bonds had previously reacted with the cellulose surface OH groups (but not all three, for steric reasons), then the remaining functionality would couple with the OH or NH_2 group of the second reagent and provide a permanent modification of the fibre surface. This strategy was successful, particularly with Et_3Al, which turned out to be much more reactive toward cellulose than Et_3B. The proofs for the validity of the approach were obtained by XPS spectroscopy and by surface energy analysis through contact angle measurements. Figure 6 shows the XPS spectrum of $AlEt_3$-treated cellulose followed by a second grafting with dicyanoamide which exhibits the characteristic peak related to aluminium and nitrogen atoms.

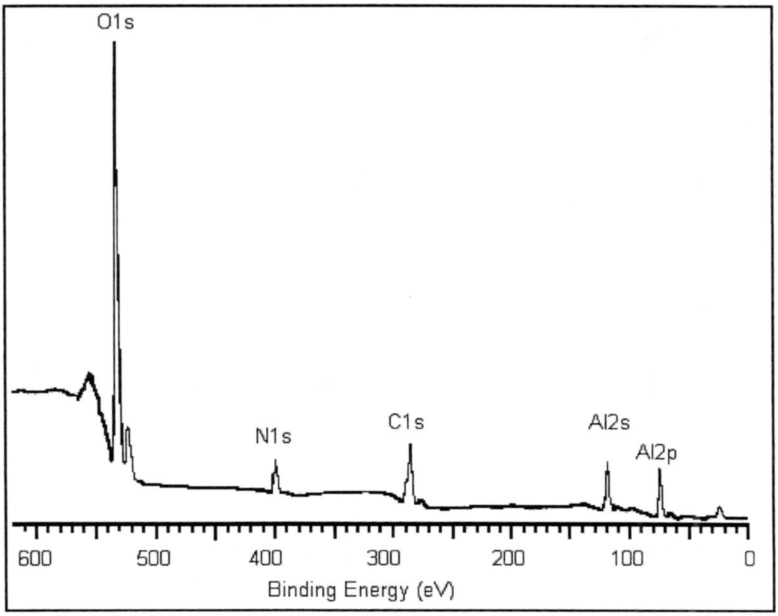

Figure 6. XPS spectrum of $AlEt_3$-treated cellulose followed by grafting with dicyanoamide in the second reaction. A Soxhlet extraction was carried out after each stage.

Conclusions

The chemical modification of the surface of cellulose fibres can be carried out following a variety of procedures and leads to interesting possibilities in terms of the type of moieties that can be appended on them. The most promising approach seems to be that giving rise to continuous covalent bonds between the cellulose surface and the matrix. Recently, we looked into a different strategy [11], which does not involve a chemical modification, but rather the formation of a physical sleeve of polymer around the fibres. This can be achieved by generating admicellar layers on the cellulose substrate with a cationic surfactant and filling it with a monomer, which is then polymerised. This type of water-based "coating" is thus a novel means of facilitating the marriage of the matrix polymer with the surface of the fibres, albeit without covalent connexion.

[1] S.J. Eichhorn, C.A. Baillie, N. Zafeiropoulos, L.Y. Mwaikamho, M.P., Ansell, A. Dufresne, K.M. Entwistle, P J., Herrera-Franco, G.C. Escamilla, L. Groom, M. Hughes, C. Hill, T G. Rials, P.M. Wild, *J Materials Science* **2001**, *36*, 2107.
[2] A. Gandini, M.N. Belgacem, *Polvm. Intern.* **1998**, *46*, 267 and refs. therein.
[3] A. Gandini, N.M. Belgacem, *J. Polymer Environment* **2002**, *10*, 105 and refs therein.
[4] J. A. Trejo-O'Reilly, J. Y. Cavaillé, N. M. Belgacem, A. Gandini, *J. Adhesion*, **1998**, 67, 359.
[5] V. Botaro, A. Gandini, *Cellulose* **1998**, *5*, 65.
[6] A. Gandini, V R. Botaro, E. Zeno, S. Bach, *Polym. Intern.*, **2001**, *50*, 7.
[7] M. Castellano, P. Fabbri, A. Gandini, M. N. Belgacem, *J. Colloid Interface Sci.*, **2004**, *273*, 505.
[8] M. Abdelmouleh, S. Boufi, A. Ben Salah, M. N. Belgacem, A. Gandini, *Langmuir* **2002**, *18*, 3203.
[9] M. Abdelmouleh, S. Boufi, A. Ben Salah, M. N. Belgacem, A. Gandini, *Intern. J. Adhesion Adhesives* **2004**, *24*, 43.
[10] P. Fabbri, Champon G., M. Castellano, M. N. Belgacem, A. Gandini, *Polvm. Intern.* **2004**, *53*, 7.
[11] S. Boufi, A. Gandini, *Cellulose* **2001**, *8*, 302.
[12] J. A. Trejo-O'Reilly, A. Gandini, J. Y. Cavaillé, *Polym. Bulletin*, **1995**, 34, 265.
[13] J. A. Trejo-O'Reilly, A. Gandini, J. Y. Cavaillé, *React. Funct. Polym.*, **1996**, 32, 9.
[14] A. Valdez-Gonzalez, J. M. Cervantes-Uc, R. Olayo, P.J. Herrera-Franco, *Composites, PartB: Engineering* **1999**, *30*, 321.

Macromol. Symp. **2005**, *221*, 271-280

Effect of Silica Particle Size on Polymer Adsorption. Morphological, Energetic and Conformational Relationships

Rami Al Akoum, Bassel Haidar, Alain Vidal*

Institut de Chimie des Surfaces et Interfaces (ICSI), CNRS-UHA, B.P. 2488, 68057 Mulhouse Cedex, France
Fax: 33 (0)3 89 60 87 99; E-mail: b.haidar@uha.fr

Summary: A series of six chemically treated and untreated fumed silicas with increasing particle size (ranging from the nano- to the micrometer size) was prepared. Surface areas (and morphologies) and surface energies were determined by nitrogen adsorption and inverse gas chromatography, respectively. The adsorption of a series of PDMS with different and well-defined molecular weights was studied at different polymer concentrations. Amounts of adsorbed polymer were determined by gravimetry and the energies of adsorption were assessed by flow-microcalorimetry. Results are discussed in terms of particle surface energy and morphology effects on the conformation and the inter-connectivity of adsorbed polymer molecules.

Keywords: adsorption; conformation; morphology; poly(dimethylsiloxane); silica; surface energy

Introduction

Polymer adsorption on solid particles is a key factor of material behavior in both colloidal and composite applications. Two measurable characteristics can theoretically define the resulting hybrids: i) the conformation of polymer chains on the solid surface, consequently the amount of adsorbed chains; ii) the strength of the polymer-solid connection. Both characteristics, for a given polymer/solid couple, are essentially governed by the particle morphology and surface energy, on the one hand, and by the affinity of the polymer towards the solid surface and consequently its adsorption energy, on the other hand. Few studies in the literature cover simultaneously these two aspects. The particular shape of the solid (finely divided particles) imposes additional specific limits on the polymer/filler organization, i.e., particle aggregation, polymer chain adsorption on multi-particles, polymer confinement, etc. The aim of the present work is to address the above-mentioned questions, the investigation part being based on several type of adsorption (physical, chemical, by condensation or by dispersive/specific interactions…). Under these conditions varying at least the two major constituents (the probe

 DOI: 10.1002/masy.200550327

and the solid surface) becomes an essential requirement. Probes should differ in their nature and size and silica of different particle diameters ought to be investigated.

Materials

Table 1. Silicas samples.

Silicas	S_0 (m^2/g)
S05	50
S13	120
S15	150
S20	190
S30	310
S40	400

Six different fumed silicas, Table 1, were provided by WACKER-CHEMIE GmbH. Their surface areas and structures were defined by controlling flame temperature and combustion time reaction.

Modification, chemical adsorption

Thanks to a rigorous control of the combustion silica synthesis, the different untreated particles turned out to present the same density of OH groups per surface unit (1.83 OH/nm^2) disregarding the actual surface area. These different silicas were progressively modified by chemical reaction with dimethyldichlorosilane (DMCS).

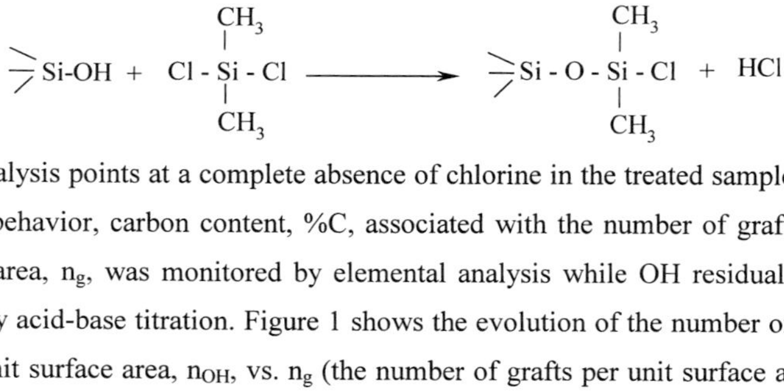

Elemental analysis points at a complete absence of chlorine in the treated samples. In order to explain this behavior, carbon content, %C, associated with the number of grafted group per unit surface area, n_g, was monitored by elemental analysis while OH residual groups were determined by acid-base titration. Figure 1 shows the evolution of the number of residual OH groups per unit surface area, n_{OH}, vs. n_g (the number of grafts per unit surface area, based on C% and the original surface area S_0). The number of grafts attached to the surface is higher than the number of disappeared OH groups, which indicates the occurrence of a substantial condensation reaction of DMCS. Thus, the complete disappearance of chlorine atoms after grafting suggests a surface reaction of the condensed chlorosilane molecule with an accessible surface hydroxyl group. In addition, we notice a similar behavior at both the molecular and surface unit levels for all studied silicas. Grafting efficiency decreases proportionally with decreasing the remaining n_{OH} irrespectively to the total OH number that is surface dependent.

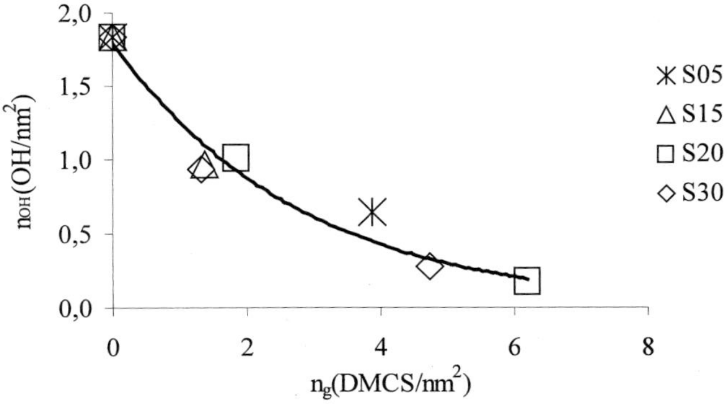

Figure 1. Grafted groups as a function of residual OH groups.

N_2 adsorption at low temperature, molecular condensation

The specific surface areas of silicas were measured before and after modification with DMCS from N_2/77K adsorption isotherms, using the BET mathematical treatment[1].

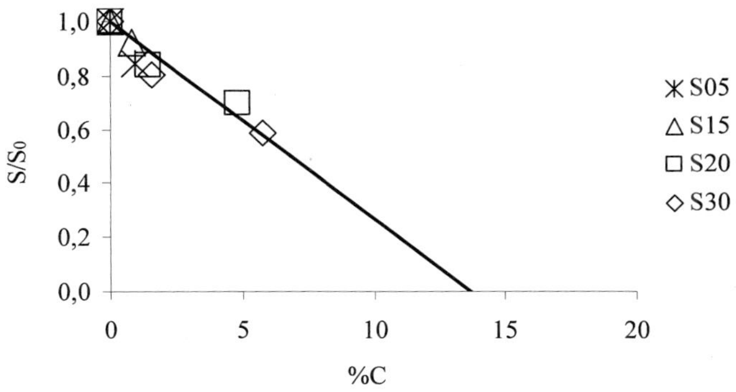

Figure 2. Evolution of reduced surface area, S/S_0, with the extent of surface modification of silicas.

Figure 2 shows that the reduced surface area (S/S_0, where S_0 refers to the untreated surface) decreases monotonically with increasing surface treatment [%C directly proportional to the total number of grafted groups; notice that C% is expressed with reference to mass unit (g) and not to the surface area unit (nm^2)]. This decrease seems to be quite similar for all silicas, i.e. one grafted group reduces the measured BET surface area by an identical percentage independently of the actual surface area. As one graft hinders or modifies the nitrogen adsorption by an extent proportional to the surface area of the silica, the higher the surface area, the lower the number of N_2 molecules which can be adsorbed on this surface after treatment. In addition, a quite good linear proportionality exists between both parameters.

Surface energy, adsorption at high temperature

Inverse Gas Chromatography (IGC) is a direct tool for the determination of the surface free energy of powders[2]. It was used in this study under infinite-dilution conditions within the 150-210°C temperature range. Results can be expressed in term of increment of free energy of adsorption, ΔG_a (related to the retention volume, *V*, of series of solutes, alkanes or siloxanes, of an increasing size) or in term of γ_s^D, dispersive component of the surface free energy. The latter is in the case of alkanes equal to[3,4]:

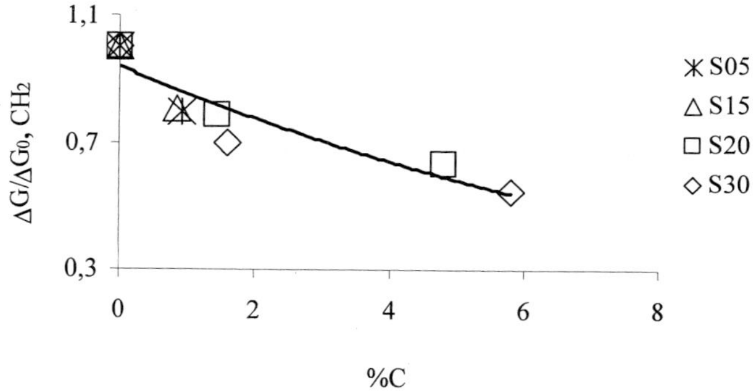

Figure 3. Evolution of the reduced free energy of adsorption of silicas with the surface modification degree (alkanes, T=160°C).

$$\Delta G_a(CH_2) = -RT.\ln\left(\frac{V_{n+1}}{V_n}\right)$$

$$\gamma_S{}^D = \left(\frac{\Delta G_a(CH_2)}{2N.aCH_2}\right)^2 \frac{1}{\gamma_{CH2}}$$

where γ_{CH2} and a_{CH2} are respectively the surface energy of a surface containing CH_2 groups only and the surface area corresponding to a methylene group.

The free (surface) energy obtained using alkane probes is shown (as reduced values or $\Delta G / \Delta G_0$, where ΔG_0 refers to the untreated silica surface) in Figure 3 as a function of the surface modification, %C, for the different silicas. ΔG_a decreases substantially with surface treatment, this reduction is similar for all silicas, whatever their surface area.

It seems that the effect of the chemical modification on the alkane free energy of adsorption goes beyond the molecular level of a graft, similarly to its effect on N_2 adsorption. Grafted molecules decrease the surface energy by an extent, which is proportional to the original surface area. This behavior is similar for all temperatures by using n-alkane probes. For siloxane probes, although we obtain qualitatively the same kind of ΔG_a reduction upon surface treatment (Figure 4), such reduction, conversely to the alkane behavior, remains dependent on the surface area (results not shown), and therefore the effect of the graft on silicas in ΔG_a is only associated with its molecular level.

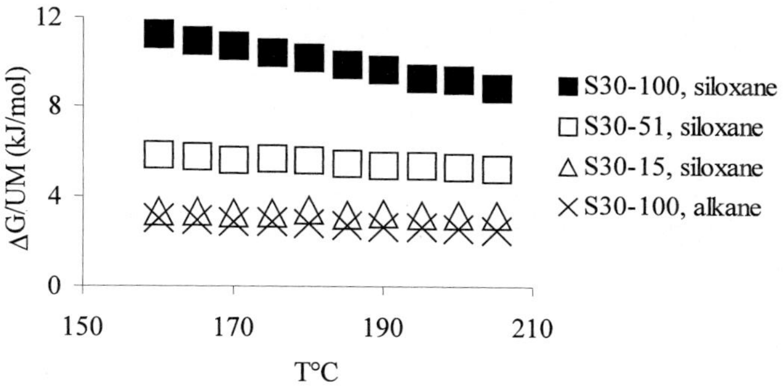

Figure 4. Siloxane free energy of adsorption-temperature dependence of untreated silica S30-100 and treated ones with 51 and 15% residual OH groups; alkanes free energy of adsorption is reported for the same untreated silica.

It is worth noticing that siloxane-silica interactions are far stronger than alkane-silica ones. Figure 4 reports ΔG_a as a function of IGC measurement temperature for untreated S30-100 and treated S30-51 and S30-15 silicas, with respectively 51 and 15% residual OH groups. ΔG_a decreases with temperature, but most of all that siloxane free energy of adsorption (thus surface energy), as well as its temperature-dependence, decreases upon treatment. Figure 4 includes for comparison $\Delta G_{a\ alkanes}$ for the untreated S30 silica.

We should notice here that even though upon surface grafting, ΔG_a siloxanes decreases with respect to ΔG_a alkane in a different way that the behavior of both probes facing modified surfaces. Furthermore, the IGC technique is highly sensitive to the surface sites of high energy and therefore could not give a "complete" picture of the actual activity for a given surface. The adsorption of polymer probes has thus to be investigated.

Flow micro calorimetry, polymer adsorption from solution

Polymer adsorption measurements, conversely to those of small molecules, need necessarily to be performed either in the melt or from solution. They can be monitored by the determination of the amount of adsorbed polymer or by the measurement of its heat (enthalpy) of adsorption. The flow microcalorimeter, FMC[5], is built to measure the small heat exchange, which occurs when molecular bonds on a surface are made or broken. It is also able of measuring the quantity of material involved in such phenomenon, thanks to an appropriate detector (UV, refractive index…).

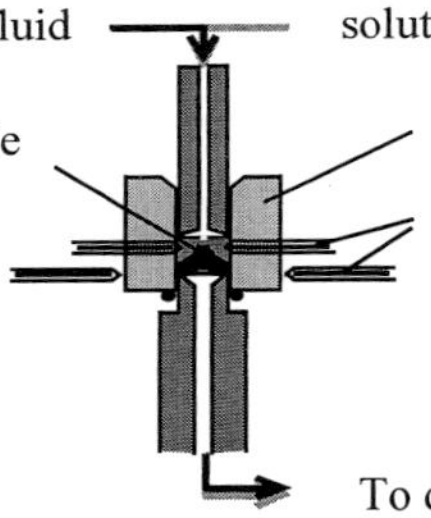

Figure 5. FMC apparatus, schematic representation.

Its fundamental principle is the following: the sample under investigation is placed in a test cell (Figure 5); a fluid (polymer solvent) is then allowed to percolate through the sample bed. All thermal events are detected thanks to four thermistors connected to a Wheatstone bridge and controlling units. Calibration is effected using a small heating coil located in the cell through which a small and controlled electric current is passed for a known time. After equilibrium is established, any modification of the fluid stream composition alters the equilibrium and the heat exchange outcome is recorded. Un-adsorbed and/or desorbed probes are detected by an appropriate downstream (refractometer or UV) detector. In this study adsorption enthalpies were determined in "pulsed mode" using a Microscal FMC with a

Perkin Elmer Totalchrom Workstation integrator. The constant flow rate was insured thanks to a Gilson 307 pump. The adsorbent was placed in the flow cell and brought in contact with n-heptane, maintained constantly under an inert atmosphere (argon). 10g/l poly(dimethylsiloxane) (PDMS) solutions were prepared using the same solvent and were introduced into the FMC, in a typical "pulsed mode" experiment, through a 20 μl loop. The actual amount of injected polymer is quite small (0.2 mg) as compared to the available large silica surface area in the cell (sample of about 100 mg). Three different PDMS (P4, P29, P420) the molecular weights (MW) of which are respectively 4000, 29000 and 420000 g/mol as well as oligomers of the same polymer such as hexamethyldisiloxane $(CH_3)_3Si\text{-}O\text{-}Si(CH_3)_3$ (HMDS), were purchased from GELEST-ABCR.

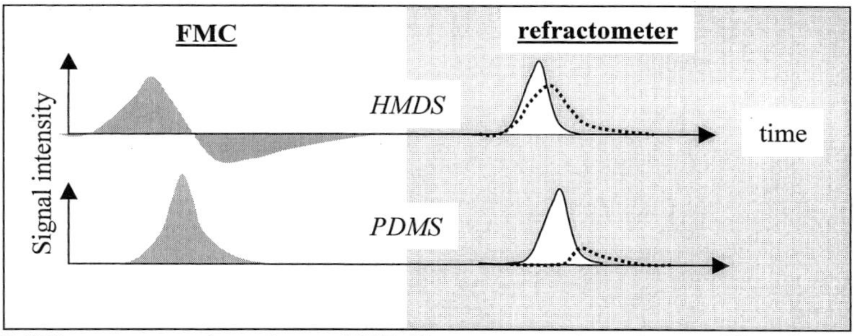

Figure 6. HMDS temporary adsorption and PDMS permanent adsorption on silica. Gray areas represent FCM thermogrammes while refractometer results are presented by continuous and dashed lines for reference injected directly in the detector and desorbed solute, respectively.

Generally in the pulsed FMC mode, low molecular weight probes (HMDS in Figure 6) show an exothermal adsorption peak upon adsorption of the solute followed by an endothermal one as the solute is desorbed from the solid surface. Almost the total amount of solute injected in the FCM through the loop gets out the cell and is detected by the downstream detector (as seen by the comparable surface of the two peaks). When a polymer such as PDMS is used as a solute it shows a peculiar behavior. Although it exchanges as much enthalpy, exothermal peak, with the silica surface as oligomers, it exhibits almost no endothermal peak (no desorption) and the fraction of retained polymer is very close to total (as supported by the small downstream detector recorded peak compared to the reference one). In this case, the adsorption is said to be permanent.

The present work offers for the first time, to our knowledge, a relatively complete and comprehensive FMC study of polymer adsorption on filler. An illustration of the usefulness of this technique is illustrated by the study of the effect of surface area on the enthalpy of adsorption. Indeed, Figure 7 shows the molar enthalpy of adsorption, E_{mol}, (the experimental heat of adsorption normalized with reference to one gram of both silica and effectively adsorbed polymer) as a function of the silica surface area for the three PDMS with different MW. It is obvious that the enthalpy involved with polymer adsorption is quite large even for the shortest chain (energy close to that of covalent bond). It gives a reasonable explanation to the permanent adsorption in the case of macromolecules in general.

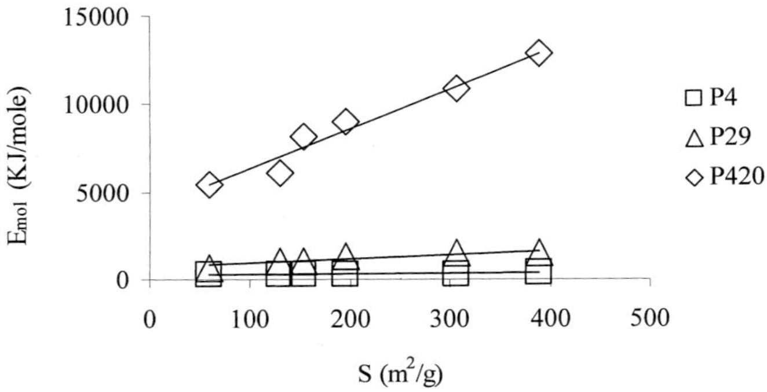

Figure 7. Molar enthalpy of adsorption, E_{mol}, as a function of surface area of silica for PDMS with different MW.

For a given molecular weight, the enthalpy of adsorption increases with the surface area of the solid. In view of the facts that (i) E_{mol} can be reasonably considered as directly proportional to the number of polymer-surface junctions; (ii) for a given molecular weight the conformation of a chain is realistically supposed to be independent of the surface area on which it may adsorb, since the only driving forces affecting the degree of spreading of a macromolecule are its molecular weight and concentration; (iii) for each molecular weight, experiments are performed under surface excess and constant concentration criteria, E_{mol} should be independent of the surface area. However, two factors may, at this point, explain the E_{mol} surface dependency. The first one is *conformational*, and is related to surface activities as expressed by surface energy, γ_s^D, which were observed to increase with increasing surface area (results not shown here but identical ones are reported by Thielmann[5]); consequently, this would rise the strength and/or the number of attractive forces which in turn would force a given chain to adopt progressively a flatter conformation.

The second is purely *geometrical* and scale-related; in fact, when dealing with closely packed powders, increasing surface area means a reduction of particle size and thus of the size of the inter-particle cavities. Consequently, for a polymer of a given molecular weight having the same conformation (random coil as a first approximation) on all surfaces, increasing surface area would lead to an increase of the number of possible contact points with the silica by multiple or inter-particle connections.

Figure 7 shows also, as expected, that the higher the molecular weight, the higher the enthalpy of adsorption. However, and in view of the above results, it is obvious that in order to compare different molecular weights and the extent of spreading of each macromolecule one should normalize the results by a conformational factor such as the gyration radius in a first approach.

In Figure 8 two kind of normalizations are made. The first one with respect to the polymer molecular weight, i.e. E_{Rg} where

$$E_{Rg} = E_{mol} / MW^{1/2}$$

$MW^{1/2}$ reflecting the gyration radius of the adsorbed molecule. The second one with respect to a state of reference, E_{Rg} / E_{Rg0}, that was taken arbitrarily as being the smallest surface, S05, of the studied silica series.

It is observed that the enthalpy of adsorption of P4 can be distinguished by a surface area-independence, conversely to what is observed for P29 and P420 which show proportionally an identical increase of the enthalpy of adsorption with increasing surface area. P4 behaviour could be explained by its eventual high and equal accessibility to all cavities without being caught up by other particles (intra-particle adsorption), given its relatively small radius of gyration. In opposition, high molecular weight adsorption seems to proceed by, essentially, an inter-particle process. It is worth noticing that the two high MW's show the same degree of inter-particles adsorption despite the significant difference in their sizes. This may be an indication that the arbitrary chosen factor of normalization is quite reasonable and not too far from reality.

Conclusion

In order to reveal surface complexity, structural, geometrical, chemical and energetic, of solids in a powder form, for instance filler particles, one should use appropriate experimental approaches specifically designed for such solids. Gas condensation at low temperature (BET) and adsorption at high temperatures (IGC) are commonly used for this purpose. FMC technique turned out to be a very valuable approach too, with the significant advantage of testing high molecular weight probes (polymers) through a fairly direct way. Furthermore, adsorption ability at variable scales (of both solid and probe) and conditions is a highly productive source of information.

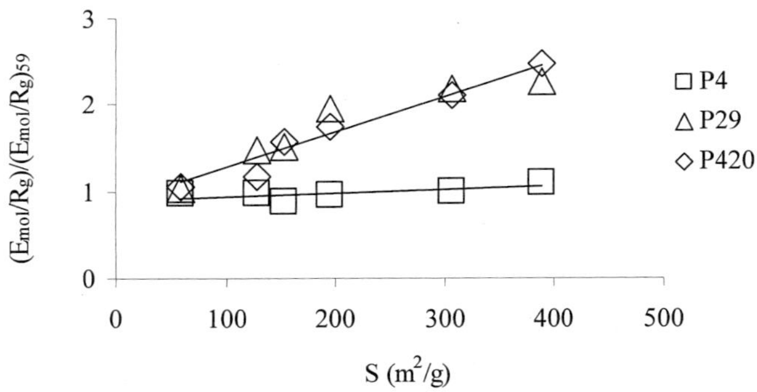

Figure 8. Reduced molar enthalpy in reference to S05 as a function of the surface area for different molecular weights.

Thanks to this approach we may underline the following conclusions:

- The effect of surface modification of silica may be both limited to the only graft size and position or overwhelming those limits to larger scale depending on the chemical nature of the probe approaching such a modified surface. Such control of chemical and energetic specificity helps out a better design of surface engineering.
- Inter-particle adsorption leads to an unusual MW-energy dependency (compared to adsorption on a flat surface or separated particles[7]; it may reveal the internal structure of filler aggregates.
- Particle packing and arrangement has a decisive role in controlling polymer-filler adsorption energy. Together with the size of the polymer (MW) it settles the adsorption nature to be of inter- or intra-particle form.

[1] S. Brunauer, P.H. Emmett, E. Teller, *J. Amer. Chem. Soc.*, **1938**, 60, 309
[2] C. Saint Flour, E. Papirer, *Ind. Engng. Prod. Res. Dev.*, **1982**, 21, 337
[3] G.D. Ulrich, *C&EN*, **1984**, 6, 22
[4] D.G. Gray et W. Hertel, *J .Appl. Poly. Sci.*, **1982**, 27,71
[5] A.J.Groszek, *Thermochimica Acta*, **1998**, 312, 133
[6] F.Thielmann, *J. Chromatography A*, **2004**, 1037, 115
[7] K.Vuillaume, B.Haidar, A.Vidal, *e-Polymers*, **2003**, no. 026